Studies in Philosophy

Outstanding Dissertations

Edited by

Robert Nozick

Pellegrino University Professor

at Harvard University

A Routledge Series

Studies in Philosophy

Robert Nozick, *General Editor*

Intuition as Evidence
Joel Pust

Teleology and the Norms of Nature
William J. Fitzpatrick

Problems of Compositionality
Zoltán Gendler Szabó

Logic and Language in Wittgenstein's *Tractatus*
Ian Proops

Thought Experiment
On the Powers and Limits of Imaginary Cases
Tamar Szabó Gendler

Between Deflationism and Correspondence Theory
Matthew McGrath

The Relevance of Phenomenology to the Philosophy of Language and Mind
Sean D. Kelly

Risk, Ambiguity and Decision
Daniel Ellsberg

The Explanationist Defense of Scientific Realism
Dorit A. Ganson

New Thoughts about Old Things
Krista Lawlor

Essays on Symmetry
Jennan Ismael

descartes' Metaphysical Reasoning

Essays on Linguistic Context Sensitivity and its Philosophical Significance
Steven Gross

Names and Nature in Plato's *Cratylus*
Rachel Barney

Reality and Impenetrability in Kant's Philosophy of Nature
Daniel Warren

Frege and the Logic of sense and Reference
Kevin C. Klement

Topics in the Philosophy of Possible Worlds
Daniel Patrick Nolan

Anthropic Bias
Observation Selection Effects
Nick Bostrom

The Beautiful Shape of the Good
Platonic and Pythagorean Theme's in Kant's Critique of the Power of Judgment
Mihaela C. Fistioc

Referential Opacity and Modal Logic
Dagfinn Føllesdal

UNDERSTANDING THE MANY

Byeong-uk Yi

Routledge
Taylor & Francis Group
New York London

Published in 2002 by
Routledge
711 Third Avenue
New York, NY 10017

Published in Great Britain by
Routledge
2 Park Square, Milton Park
Abingdon, Oxfordshire OX14 4RN

First issued in paperback 2014

Routledge is an imprint of the Taylor & Francis Group, an informa business

Library of Congress Cataloging-in-Publication Data

Yi, Byeong-uk.
Understanding the many / Byeong-uk Yi.
p. cm. — (Studies in philosophy)
Includes bibliographical references (p.) and index.
ISBN 978-0-415-93864-8 (hbk)
ISBN 978-0-415-86688-0 (pbk)
1. Set theory—Philosophy. 2. Pluralism. I. Title. II. Studies in philosophy (New York, N.Y.)
QA248.Y5 2002
511.3'22—dc22 2002017874

TO MY PARENTS

Contents

Preface

What follows is essentially the dissertation that I submitted for a Ph.D. in Philosophy at UCLA in 1995.[1] The dissertation contains an interim report of my investigations into the nature of *many things as such* and the expressions that we use to talk about them, such as the plural constructions of natural languages.

The investigations stemmed from an attempt at understanding the nature of sets. According to the standard conception of set, a set (e.g., the set {Russell, Whitehead}) is something determined by its members (e.g., the two humans: Russell and Whitehead). While attempting to clarify and defend this idea, I came to think that drawing a sharp and unmitigated distinction between a set and its members is crucial to the proper understanding, let alone defense, of the idea. The distinction, in the typical cases in which the set in question has more than one member, boils down to the distinction between the *one* and the *many*, between some *one thing* and the *many things* that are somehow related to it: the set {Russell, Whitehead} is to be distinguished from Russell and Whitehead (not only from each one of them), because the set is only one thing while the humans

1. For the *Studies in Philosophy* series I have made some stylistic changes. There remains much room for improvement, but it would have taken me far afield to present its content as I would now do. Despite various infelicities in presentation, I am inclined to think that most of what I claim in it is defensible. For more recent presentations of some of its themes, see "Numbers and Relations", *Erkenntnis*, 49 (1) (1998): 93-113; "Is Mereology Ontologically Innocent?", *Philosophical Studies*, 93 (2) (Feb., 1999): 141-60; "Is Two a Property?", *Journal of Philosophy*, 96 (4) (Apr., 1999): 163-90; "Is There a Plural Object?", *Studia Logica*, forthcoming; and "The Language and Logic of Plurals", *Journal of Philosophical Logic*, forthcoming.

are many things (viz. two humans). Would anyone ignore or attempt to blur or mitigate the obvious distinction? Some might think that the two humans, conceived as one thing, are a set and that the set is just the two humans as so conceived, which means that the set, which is one thing, must also be many things. Those who would not go so far as to hold that the set *is* the two humans might still hold that the set is not only one thing (e.g., one set) but also many things (e.g., two humans). And even those who keep the distinction quite clear might still think that the existence or nature of the two humans *as such* (but not that of either one of them) rests on the existence or nature of the set, some one thing that comprehends them. Or they might think that we cannot talk or think about the two humans as such without talking or thinking, if only implicitly, about the set. I think that these are confused ideas that yield distorted pictures of the nature of the many things as such, let alone the nature of the set, and that keep us from the proper understanding of our talks and thoughts about them.

One might well *conceive* of two humans, such as Russell and Whitehead, as some one thing, be it a set or not. But what one will then get is simply a misconception about the humans. The wrong conception does not make the humans themselves the one thing, which they are not identical with prior to the conception.[2] Now some might say that the two humans are identical with some one thing that comprehends them (call it *Genie*)–whether or not they are taken to be one. But this is not correct; Genie is one of *Genie*, but Genie is not one of *Russell and Whitehead*.[3] And it is hard to see how one can be led to think that a

---

2. This is not to deny that one might have the power to turn the humans into some one thing by merely conceiving them as that, but their turning into that one thing does not mean that they *are* that one thing. A Harry Potter might turn a human into a cat simply by conceiving her as a cat, but this does not mean that the human, as conceived as a cat by him, *is* the cat that she is transformed to; despite its power to cause a substantial change, his conception would be a wrong conception of the human. This, to be sure, does not mean that there can be no animals that, spontaneously or with the aid of a Harry Potter, can be like humans during the day but like cats at night; such animals (e.g., cat-people) would be neither humans nor cats–given what humans and cats are like.

3. Genie is not one of Russell and Whitehead, because Genie can be neither Russell nor Whitehead (and this because Russell is not Whitehead). Those who have qualms about the argument because of the seeming grammatical anomaly of "Genie is one of *Genie*" can consider the following: Genie is one of [*Genie* and Cicero], but not one of [*Russell and Whitehead* and Cicero]. For more

set can *be* many things, as well as one thing, without identifying a set mistakenly with its members. Surely, the set {Russell, Whitehead}, for example, *is composed of* many things (viz. two humans), but this does not make the set itself many things any more than Chelsea's being an offspring of two adults makes Chelsea herself two adults.

Does the existence or nature of the two humans depend on the existence or nature of the set? I think not. Russell and Whitehead exist as two humans, as long as Russell exists as a human, Whitehead exists as a human, and Russell is not Whitehead.[4] Some might grant this but still hold that to *talk* about the many things as such must be to talk, if implicitly, about the set. On their view, to say, for example, "Russell is one of the philosophers who admire only one another" is to commit oneself to the existence of some one thing (e.g., a set) that comprehends many philosophers (e.g., Russell and Whitehead). But to say that is not at all to commit oneself to the existence of any such thing. The sentence is logically implied by "Russell is one of Russell and Whitehead, and Russell and Whitehead are the philosophers who admire only one another", which is in turn logically implied by "Russell is Russell or Whitehead, Russell admires only Whitehead, Whitehead admires only Russell, and Russell is not Whitehead", which does not logically imply the existence of any such thing.

The attempts to understand the nature of the many and our talks about them by invoking some one thing that comprehends them can be seen to stem from a longstanding *bias for singularity*, a bias that permeates contemporary understandings of language, thought, and reality.[5] The bias manifests itself as a reductionist thesis about plurals: the plural constructions of natural languages (e.g., the plural term "Russell and Whitehead" or the predicate "are two humans" of the plural form) must be reducible to the singular constructions (e.g., the singular term "Russell" or the predicate "is a human" of the singular form). This thesis about language rests ultimately on the standard conception of

---

discussions on the argument, see Yi, "Is Mereology Ontologically Innocent?", *Philosophical Studies*, 93 (2) (Feb., 1999): 141-60; and Yi, "Is There a Plural Object?", *Studia Logica*, forthcoming.

4. I think that the ontological dependence runs the other way around: the set exists because its members do, as on the standard conception of set, and its *being two-membered* rests on its members *being two things*.

5. The bias can be traced back to Aristotle. See Aristotle, *Categories*, translated by J. L. Ackrill in *Categories and De Interpretatione* (Oxford: Clarendon Press, 1963).

reality,[6] according to which *many* things *as such* cannot have a property.[7] On this conception, Russell and Whitehead, for example, can have no nature of their own–although each one of them can have a nature by having, e.g., the property of being a human. There could be no such property as *being two humans*, a property signified by the predicate "to be two humans", on the conception, because it would be a property instantiated only by *two* things as such, not by either one of them. If so, we cannot be referring to the two humans to attribute a property to them when we say "Russell and Whitehead are two humans", "They are many philosophers who admire only another", and so on. And one might conclude from this that those who say these must be attributing a property to some one thing that comprehends the humans, such as a two-membered set, and be led to identify the one thing with the two humans or else to think that their being two humans rests on its being two-membered.

The dissertation reprinted below contains the results of my first attempts at developing a framework for understanding the many as such and our talks about them by redressing the prevalent and longstanding bias at its roots. I reject the thesis that many things as such cannot instantiate a property and propose to replace the standard conception of reality, which incorporates the thesis, with a more liberal conception. On this conception, which can be called the *plural conception of reality*, there are *plural properties* that can be instantiated by *many things as such. Being two humans* (or, for that matter, *being many things*), for example, is a property instantiated by Russell and Whitehead as such, not by either of them.[8] Accordingly, I argue that plural constructions cannot be reduced to singular constructions. They must be understood on their own–plural terms (e.g., "Russell and Whitehead") as referring to many things (e.g., Russell and Whitehead) as such, and the 'plural predicates' (e.g., "to be two humans") as

6. For the classic sources of the conception, see Aristotle, *ibid.*; and Gottlob Frege, "Über Begriff und Gegenstand", *Zeitschrift für Philosophie und Philosophische Kritik*, 100 (1895): 25-50 (this article, together with its preprint version that contains additional discussions of plural constructions, is translated as "On concept and object" in Frege, *Posthumous Writings*, Hans Hermes et. al. (ed.) (Oxford: Basil Blackwell, 1979): 87-117).

7. Or, in general, many things as such cannot fill in one argument place of a relation.

8. And there are *plural relations*, such as *coauthoring*, which Russell and Whitehead as such have in relation to *Principia Mathematica*.

signifying plural properties or relations.[9] To develop the anti-reductionist account, I construct *regimented* 1st-order *languages* that extend elementary languages by including the plural counterparts of resources available in elementary languages (viz. plural terms and predicates, and plural quantifiers and variables) while presenting a scheme of *paraphrasing the plural constructions* of natural languages into the extended languages, and present a *logical system*, a conservative extension of elementary logic that I call the *logic of plurals*, that helps to clarify the logical relations among plural, as well as singular, constructions.

The framework that I have developed for understanding the many casts light on the nature of numbers as well as the nature of sets. The plural conception of reality, aligned with the anti-reductionist treatment of plurals, allows one to give a natural analysis of various facts about natural numbers and our talks about them (e.g., "the number of the authors of *Principia Mathematica* is 2"). I argue that contemporary conceptions of natural numbers are based on wrong analyses of numerical facts that rest on the standard conception of reality and that their natural analysis yields the view that *natural numbers* (e.g., 2) are *plural properties* (e.g., being two things). And my interests in the nature of sets are reflected in the last chapter of the dissertation, where I attempt to explain the way in which a set is determined by its members by relating it directly to its members without mediation of anything else, be it the set itself or the collection or aggregate that consists of the members.[10]

---

9. For example, the predicate "to coauthor" in "Russell and Whitehead coauthor *Principia Mathematica*" is a two-place plural predicate that signifies the two-place relation of coauthoring.

10. Many people helped me while I was working on these themes during my graduate study at UCLA and thereafter. I wish to thank them all for their teachings and encouragements, criticisms and advice. In addition to those named below in the beginning of the dissertation, I must mention Bob Hale, Bernie Linsky, Jeff Pelletier, and Graham Priest, who helped me to improve my understanding of the many. I would also like to thank Robert Nozick, the editor of the *Studies in Philosophy* series, Damian Treffs, Routledge's Dissertations editor, and Kendall Huffhines and Chung-Hyoung Lee, who helped me to get the final version of this book. And special thanks are due, again, to Tony Martin.

Acknowledgments

Tony Anderson, Tim Bays, Tyler Burge, Andrew Hsu, Amy Kind, and Karel Lambert gave me valuable comments, criticisms, and encouragement while I was working on this monograph. Glenn Branch, Stavroula Glezakos, and Daniel Krasner read and commented on the penultimate version. The works and advice of Joseph Almog, David Kaplan, Kit Fine, Penelope Maddy, and Yiannis Moschovakis were critical to the shaping of my thoughts. I am especially indebted to my supervisor Tony Martin, who guided me through all the stages of writing this work. And I wish to thank Eunshil Bae, whose support enabled me to launch and complete this work.

Introduction

This monograph contains five chapters. They are written as separate papers that address various related issues. The unity underlying their diversity is the pursuit of a common understanding of what are *many* things or, more precisely, what are *more* numerous *than one* thing.

Russell, for example, is one thing; he is a human being, a philosopher. It is the same with Whitehead. But Russell *and* Whitehead (taken together) are more than one thing. They are two different philosophers, because Russell is not identical with Whitehead.

This distinction naturally leads to the view that characterizes my understanding of what are many (i.e., more than one): There are properties instantiated by what are more than one thing (as such) as well as properties instantiated by what is one thing. Being a philosopher is a property instantiated by Russell; Russell *is a philosopher*. Similarly, I maintain, *being more than one philosopher* is a property instantiated by Russell and Whitehead (as such);[1] they are more than one philosopher (viz. two philosophers). *Being two things* (e.g., two philosophers) is also such a property; Russell and Whitehead are two things (e.g., two philosophers).[2]

1. This, to be sure, is not to say that being more than one thing is instantiated by Russell and also by Whitehead, any more than to say that Russell and Whitehead are more numerous than one thing is to say that Russell, as well as Whitehead, is more numerous than one thing.

2. Similarly, what are many (as such) have a relation to something (or some things). Russell and Whitehead have the relation of being-the-authors-of to *Principia Mathematica*, although neither of them has the same relation to the book.

In Chapter 3, the central chapter of this monograph, I aim to clarify and defend this view, which seems to be almost universally rejected as a non-starter.[3] First, I clarify the content of the view by relating it to the issue of giving the proper analysis of, e.g., the following fact:

I. Russell and Whitehead are two things.

Then I argue that such facts must be analyzed in terms of what are many things (e.g., Russell and Whitehead) and a property that they, as such, instantiate (e.g., being two things).[4]

My argument against a prominent alternative to this analysis of fact *I* brings in logical considerations. On the alternative analysis, fact *I* is identified with a fact that contains the property of *being two-membered* instead of the property of being two things:[5]

II. The set {Russell, Whitehead} is two-membered.

On this analysis, the following must be a logical consequence of fact *I*:

III. There is something (e.g., {Russell, Whitehead}) of which something else (e.g., Russell) is a member.

For this fact is a logical consequence of fact *II*. Thus the analysis is wrong because, I hold, fact *III* is not a logical consequence of fact *I*.

My view that fact *III* is not a logical consequence of fact *I* is closely related to the view that the following is not a logically valid argument:

Russell and Whitehead are two things.

---

3. Or it seems to be confused with or assimilated to different views considered more plausible, e.g., the view that being many (or being two) is instantiated by some one thing that in a way comprehends the many things in question, e.g., the set {Russell, Whitehead} or the aggregate consisting of Russell and Whitehead.

4. There I call such a property a plural (or, plurally instantiated) property.

5. Thus the analysis does away with a property instantiated by what are many things; being two-membered is not instantiated by what are many (as such) but by some one thing that has many members, e.g., the set {Russell, Whitehead}.

Therefore, there is something (e.g., {Russell, Whitehead}) of which something else (e.g., Russell) is a member.

I devote Chapter 1 to arguing that this and other similar arguments, such as the following, are not logically valid:[6]

Some critics (e.g., Ezra and Thomas) admire only one another.
Therefore, there is something (e.g., {Ezra, Thomas}) each member of which admires something only if the latter is another member of the former.

The case against the validity of such arguments appeals to the intuition that the following, for example, are logically valid arguments:

Russell is a thing, Whitehead is a thing, and Russell is not identical with Whitehead.
Therefore, Russell and Whitehead are two things.

Ezra and Thomas are critics who admire only each other.
Therefore, some critics admire only one another.[7]

It should also be clear that the first of the following two arguments is logically valid whereas the second is not:

Russell and Whitehead are two things.
Therefore, there are some things that are two things.

Russell and Whitehead are two things.

---

6. And I use the result to refute the view, proposed by, e.g., W. V. Quine, that the conclusion of the second argument mentioned above is an analysis or paraphrase of its premiss "Some critics admire only one another."

7. In addition, the following, for example, is not logically valid:

Russell is a thing, Whitehead is a thing, and Russell is not identical with Whitehead.
Therefore, there is something (e.g., {Russell, Whitehead}) of which something else (e.g., Russell) is a member.

Therefore, there is something that is two things.

In Chapter 2, I present a conservative extension of elementary logic that does justice to the intuitions.

The last two chapters, Chapters 4 and 5, apply the framework expounded in the earlier chapters. Chapter 4 is directly connected to Chapter 3, where it is argued that there are numerical properties corresponding to natural numbers: being one thing, being two things, being three things, and so on.[8] In Chapter 4, I identify natural numbers as those numerical properties: the number 2, for example, is the property of being two things. I argue that arithmetical facts, such as the following, must also be analyzed in terms of the numerical properties featuring in the analyses of ordinary numerical facts, such as *I*:

IV. Two plus one is three (in symbols, 2+1=3).

Then I identify the numerical properties, the common components of the numerical facts of the two kinds, as natural numbers.

In Chapter 5, I show how my understanding of what are many things is relevant to understanding the nature of sets. On my view, the existence or nature of what are many (e.g., Russell and Whitehead) does not presuppose that there is some one thing that comprehends them, e.g., the set {Russell, Whitehead}. This understanding of the many is essential to doing justice to a core idea of the standard conception of set: a set is something determined by its members. In Chapter 5, I explain a way in which a set is so determined by relating it directly to its members without the mediation of anything else, such as the set itself or even the collection or aggregate that consists of them. This helps to justify, as well as clarify, the standard conception of set.

8. Notice that most of the properties are instantiated by what are many things (as such). Being one thing is an exception.

1. Plural Quantifications

In *Roots of Reference*, Quine says:

> Occasionally . . . [people] say things that involve no explicit mention of classes or attributes but that come nevertheless to call for quantification over classes when we paraphrase them into economically schematic language. (1973, 111)

As a good case of this, he refers to the Geach-Kaplan sentence:

(1) Some critics admire only one another.

On his view, the sentence can be paraphrased as follows:[1]

(2) $\exists a(\exists x x \varepsilon a \wedge \forall x[x \varepsilon a \rightarrow x$ is-a-critic $\wedge \forall y(x$ admires $y \rightarrow x \neq y \wedge y \varepsilon a)])$

This sentence contains a quantifier (viz., "$\exists a$") the range of which must include some sets (or classes), such as sets of critics.

The view that sentence (1) can be paraphrased by sentence (2) has direct bearing on the logical status of, for example, the following argument:

> Some critics admire only one another.
> Therefore, there is something of which a critic is a member.

1. See Quine (1974, 239). I have made obvious notational changes. See also Quine (1972, 239) and Quine (1982, 293).

The conclusion of this argument, i.e.,

(3) There is something of which a critic is a member.

can be paraphrased as follows:

(4) $\exists a \exists x(x \text{ is-a-critic} \wedge x \varepsilon a)$.

Sentence (2) logically implies sentence (4)—on elementary logic. Thus on Quine's view, sentence (1) logically implies sentence (3) and the above argument is logically valid.[2]

My aim in this chapter is to show that the Geach-Kaplan sentence (1) does not logically imply sentence (3) and that sentence (2) is not a correct paraphrase of sentence (1). In section 1, I examine some considerations that might motivate the view that (1) can be paraphrased by (2) and explain that they fail to support the view. In section 2, I argue that (1) does not imply (3). In section 3, I use the result of section 2 to argue that (1) cannot be paraphrased by (2).

1.1 Singular Versus Plural Quantifications

The Geach-Kaplan sentence "Some critics admire only one another" can be paraphrased as follows:

(5) *There are some things such that* anything that *is one of them* is a critic and admires something only if it is another thing that *is one of them.*

---

2. This is closely related to the view that asserting the Geach-Kaplan sentence commits one to sets (or classes) of critics. Thus Quine says "The serious motive for assuming [the existence of] classes is to be found in constructions such as that of" the Geach-Kaplan sentence (1974, 239f). Most, if not all, of the issues usually put in terms of the so-called ontological commitments of a sentence (or its assertion) can be put in terms of whether the sentence in question (e.g., the Geach-Kaplan sentence) logically implies such sentences as (3).

By paraphrasing away "one another" in the Geach-Kaplan sentence, we get "There are some critics *each of* whom admires only *another one of them*", which can be rephrased as "There are some things such that *each of* them is a critic and admires only *another one of* them." Replacing "*each of* them" and "admires only *another one of* them" in this sentence with "anything that is one of them" and "admires something only if it is another one of them", respectively, yields sentence (5).

Now, sentence (2) can be seen to result from an application of a scheme of singularization to sentence (5). This seems to be a major consideration that leads to the view that the Geach-Kaplan sentence can be paraphrased by sentence (2) and so involves an implicit quantification over sets (or classes) of critics.[3]

First, replace the predicate "is one of" in sentence (5) with "is a member of", which is a predicate indicating the membership relation that holds between a member (e.g., Ezra or Thomas) of a set or class (e.g., {Ezra, Thomas}) and the set or class. Second, replace accordingly the plural quantifier construction "There are some things such that . . ." and the associated pronoun "them" in sentence (5) with the singular quantifier construction "There is something such

3. It is usual to paraphrase the Geach-Kaplan sentence (1) by a second-order sentence, to which the natural paraphrase (5) of (1) tends to be assimilated:

$$\exists F(\exists xFx \land \forall x[Fx \rightarrow x \text{ is-a-critic} \land \forall y(x \text{ admires } y \rightarrow x{\neq}y{\land}Fy)]).$$

This practice suggests the idea that the syntax or semantics of sentence (1) or (5) must be understood by way of the second-order sentence: the plural quantifier "There are some things such that" and the plural pronoun "them" (or else the phrase "is one of them") in (5), for example, are the second-order quantifier and the associated variable (e.g., "*F*" or their English analogues), respectively. I reject this idea and distinguish sentence (1) or (5) sharply from the alleged second-order paraphrase. (But it is another question whether the second-order sentence (or "formula") can be *re*-interpreted (or interpreted) to mean what sentence (1) or (5) means, but not what sentence (3) means.) In Chapter 2, I develop another regimentation of, e.g., English that can render natural paraphrases of English plural constructions (e.g., (1) or (5)), and an alternative to second-order logic that can do justice to such sentences.

that [*there is something that is a member of the former and . . .*]"[4] and the singular pronoun "it", respectively. Then we get:

> (6) There is something such that [there is something that is a member of the former and anything that is a member of it is a critic, and admires something only if it is another thing that is a member of it].[5]

Sentence (2), in which "ε" is a symbolic abbreviation of the predicate "is a member of" for membership, is a straightforward paraphrase of this sentence into *elementary notation*,[6] the canonical notation developed with elementary logic.

Another major consideration that lies behind the alleged paraphrase (2) is David Kaplan's result about the Geach-Kaplan sentence. In *Methods of Logic*, Quine explains the result before offering sentence (2) as its paraphrase:

> Kaplan has proved that we cannot express [the sentence] in terms of just ['critic'], 'admire', identity, truth functions, and quantification . . . . Invoking classes and membership, we can do justice to [the sentence]. (1982, 293)

That is, no sentence in elementary notation that contains at most the three predicates "is identical with" (in symbols, "="), "is a critic," and "admires" (in

---

4. It is essential to add the italicized phrase. Otherwise the sentence we get does not logically imply "There is a critic", which sentence (5) clearly does.

5. Michael Resnik seems to support the scheme of singularization; he says that sentence (5), for example, "seems to me to refer to collections quite explicitly. How else are we to understand the phrase 'one of them' other than as referring to some collection and as saying that the referent of 'one' belongs to it?" (1988, 77).

6. For the resources available in elementary notation, see Chapter 2, section 2.1.1. It is usual to call elementary logic and elementary notation *first-order logic* and *first-order notation*, respectively. I avoid "first-order logic" and "first-order notation" because they suggest contrasts with second-order logic and second-order notation as their natural extensions. The extensions of elementary notation and elementary logic that I develop in Chapter 2 are their first-order extensions.

addition to the basic resources of the notation: truth-functional connectives, the quantifiers "∃" and "∀", and the variables "*x*", "*y*", etc.) is a correct paraphrase of the sentence.[7]

But the most we can get out of Kaplan's result is that *if* there is a sentence in elementary notation that paraphrases the Geach-Kaplan sentence, it must contain an additional predicate. This gives no reason to take the sentence to be paraphrased by sentence (2) or, for that matter, any other sentence in elementary notation; there is the possibility that the Geach-Kaplan sentence cannot be paraphrased into elementary notation at all.

The first consideration, the one that appeals to the singularization scheme, also provides a poor case for the alleged paraphrase. The scheme used to singularize (5) does not in general yield right results. Consider, for example, "Some critics live in New York." This sentence can be paraphrased as "There are some critics each of whom lives in New York" and then as "There are some things such that anything that *is one of* them is a critic and lives in New York." Now, the scheme renders this sentence to "There is something such that [there is something that is a member of the former and anything that is a member of it is a critic and lives in New York]" (colloquially, "There is something, with a member, each member of which is a critic that lives in New York"). But it is one thing to say "Some critics live in New York", quite another to say "There is something, with a member, each member of which is a critic that lives in New York." In particular, "Ezra and Thomas are critics that live in New York" (which is logically implied by "Ezra is a critic that lives in New York and Thomas is a critic that lives in New York") logically implies the former sentence but not the latter.[8]

---

7. The point of the proof is in effect to consider a variation of the sentence: "Some positive natural numbers succeed only one another", where number n is said to *succeed* number m if and only if $n=m+1$. With the aid of this variation of the Geach-Kaplan sentence the existence of non-standard integers could be expressed. For an account of Kaplan's proof, see Boolos (1984, 432f).

8. For another example, the scheme renders "Some critics live in New York and Ezra *is one of* them" as "There is something with a member, each member of it is a critic that lives in New York, and Ezra is a member of it." Notice also that "(is) one of" has an obvious connection to "(are) some of". So the scheme renders "Some critics live in New York and some of them are British" as, roughly, "There is something x with a member, each member of x is a critic that lives in New York, and there is something y with a member such that each

The two considerations, we have seen, fail to support Quine's view that sentence (2) paraphrases the Geach-Kaplan sentence. Accordingly, the view has been subject to serious challenges by, most notably, George Boolos and David Lewis.[9] Still it seems fair to say that Quine's is the prevailing view about the Geach-Kaplan sentence. I think this is due in part to the fact that a convincing argument that directly refutes the view has not been given yet.[10] I aim to present such an argument in the next two sections.

---

member of *y* is a member of *x* and is a British." These results, I think, are not acceptable, either.

9. See Boolos (1984) and (1985a), and Lewis (1991). They are criticized by Resnik (1988) and Hazen (1993).

10. To argue against the view, Boolos and Lewis attempt to refute a sweeping thesis about all plural constructions: "the idea that use of plural forms must always be understood to commit one to the existence of sets (or 'classes,' 'collections,' or 'totalities') of those things to which the corresponding singular forms apply" (Boolos 1984, 442) or the "dogma" that "Plurals . . . are the means whereby ordinary language talks about classes" (Lewis 1991, 65). A strong case against the 'dogma' can be found in, for example, "Some critics live in New York"; but to refute the 'dogma' is not to refute the view in question but only to refute a ground on which it might be supported. Moreover, the 'dogma' does not seem to be a ground on which Quine supports his view: otherwise he would not have needed to depend on Kaplan's result to find cases of implicit quantification over sets (or classes); "Some critics live in New York" (or "There *are* centaur*s*" or "There *are* prime number*s* larger than one million") would have worked just as well. Resnik also distinguishes 'genuine' plural quantifications from merely apparent ones that "can be plausibly put into [elementary notation]" (1988, 80). To be fair to Boolos and Lewis, they aim at more than just refuting the 'dogma'; in arguing against it, they consider sentences that seem to be closer to the Geach-Kaplan sentence than the humdrum plural constructions mentioned above: e.g., "There are some sets of which every set that is not a member of itself is one" (Boolos, *ibid.*) or "There are the non-self-membered classes" (Lewis, *ibid.*). Still, they have yet to show why the semantics of the Geach-Kaplan sentence must be tied to that of these sentences while the semantics of the humdrum ones might not.

1.2 Implication

In this section, I argue that the Geach-Kaplan sentence "Some critics admire only one another" does not logically imply "There is something of which a critic is a member."

1.2.1 The Implication Argument

First, consider the following sentences:

(7) Ezra is a critic, and Thomas is a critic; Ezra is not identical with Thomas; Ezra admires only Thomas, and Thomas admires only Ezra.

(8) Ezra and Thomas are [two different] critics who admire only each other.

(1) Some critics admire only one another.

It should be clear that

A. Sentence (7) logically implies sentence (8).

B. Sentence (8) logically implies sentence (1).

Consequently, sentence (7) logically implies the Geach-Kaplan sentence (1). Second, consider sentences (7) and the following:

(3) There is something of which a critic is a member.

I think it should also be clear that

C. Sentence (7) does not logically imply sentence (3).

Finally, we can combine the two results: sentence (7) logically implies sentence (1), but not sentence (3). Then it follows that the Geach-Kaplan sentence (1) does *not* logically imply sentence (3). This completes my argument (call it the *Implication Argument*).

1.2.2 On the Three Theses on Implications

I put Theses *A-C* and their result in terms of the notion of *logical implication*. This notion is tied to the notion of *logical validity*: sentence φ logically implies sentence ψ, if the argument whose premiss is φ while its conclusion is ψ is logically valid. Thus Thesis *A* is in effect the thesis that the following is a logically valid argument:

> Ezra is a critic, and Thomas is a critic; Ezra is not identical with Thomas; Ezra admires only Thomas, and Thomas admires only Ezra.
> Therefore, Ezra and Thomas are [two different] critics that admire only each other.

Similarly, the other two theses and their result can also be put in terms of the notion of logical validity. Now, the conclusion of the Implication Argument directly follows from the three theses and the theses have strong intuitive appeal. But it is useful to elaborate on the nature and strength of the appeal.

We have firm intuitions regarding logical relations among some, if not all, sentences of natural languages, such as English. We have a clear grasp that "Ezra is a critic and Thomas is a critic" logically implies "Ezra and Thomas are critics" (and also *vice versa*). We can likewise see that "Ezra is a critic who admires only another critic" logically implies "There is a critic who admires only another critic" (but that it does not logically imply "Thomas is a critic who admires only another critic"). I maintain that we can see, in the same way, that Theses *A* and *B* hold.

This is not to say, to be sure, that we can do no more than simply appeal to the intuitions to support them. We can and do reflect on intuitions regarding implication relations among, e.g., English sentences, compare and weigh the intuitions, and sometimes construct a more or less elaborate theory that systematizes and accounts for the correctness of most, if not all, of the intuitions. Thus, on reflection, the correctness of Thesis *A* has much to do with the "each other" construction of sentence (8), "Ezra and Thomas are critics who admire only *each other*." And the correctness of Thesis *B*, that sentence (8) logically implies the Geach-Kaplan sentence, must be explained in terms of the function of the plural quantifier "some" in the sentence.[11] Here, however, I avoid simply presupposing the correctness of such a theory. The more or less elaborate accounts, such as those suggested above of the "each other" construction and the plural quantifier "some", themselves require defense, and their defense must in the end rely on such intuitions as those adduced for the two theses *A* and *B* that are here in question. It is especially so with my account of the plural quantifier in the Geach-Kaplan sentence; the account is a competitor of the Quinean account implicit in the view that sentence (2) paraphrases it. It is partly in order to refute the Quinean account that I advance the Implication Argument and dispute the alleged paraphrase. So the Implication Argument cannot presuppose the alternative to the Quinean account.

And there is no need to presuppose the correctness of my accounts. The intuitive supports for Theses *A* and *B* are stronger than those that such accounts can provide. They are, I maintain, strong enough to work as fairly stable data that can be used in the Implication Argument and, in the end, evaluating competing accounts of the key constructions.

It remains to consider Thesis *C*. According to the thesis, sentence (7) does not logically imply sentence (3); that is, the following argument is not logically valid:

> Ezra is a critic, and Thomas is a critic; Ezra is not identical with Thomas; Ezra admires only Thomas, and Thomas admires only Ezra.
> Therefore, there is something of which a critic is a member.

I think one can recognize this argument as invalid in a way comparable to the way one can see that the following arguments are so:

> Ezra is a critic.
> Therefore, there is something of which Ezra is a member.

> Chelsea lives in Washington.
> Therefore, there is something that is a mother of Chelsea.

11. For example, "John and Carol are children who carry a piano upstairs" likewise logically implies the plural quantification "Some children carry a piano upstairs."

The rough idea underlying the assessments of the three arguments can be put as follows. Whether or not Chelsea lives in Washington is a matter independent of whether she has a mother; similarly, whether Ezra is a critic or not is independent of whether he is a member of something (e.g., a set or class). So the second and third of the three arguments are invalid. Then the first one, which is in question in Thesis *C*, must also be invalid; whether Ezra and Thomas are critics, whom they admire, and whether they are the same person or not must all be likewise independent of whether Ezra or Thomas (or another critic) is a member of something (e.g., a set or class). It is beyond the scope of this chapter to make this idea precise and, in particular, to clarify the notion of independence here in question. Suffice it to say that I think there is a way in which elementary logic embodies the idea.[12] In any case, it is worthwhile to notice that elementary logic verifies the assessment of the argument in question in Thesis *C* (as well as those of the other two arguments mentioned above). The argument can be formulated in elementary notation; its premiss and conclusion (i.e., sentences (7) and (3)) can be straightforwardly paraphrased into the notation. And the argument, so formulated, is invalid on elementary logic. Thus one who rejects Thesis *C* to challenge the Implication Argument must reject elementary logic.

1.2.3 Implication Versus Metaphysical Necessity

Nevertheless, the main, or at least the most prevalent, objection that I have heard made against the Implication Argument is one that challenges Thesis *C*. The objection is based on specifying the usual characterization of the logical implication relation in terms of the recent notion of metaphysical necessity.

It is usual to characterize the logical implication relation in terms of modality: Sentence φ *logically implies* sentence ψ if, and only if, it is necessary that if φ then ψ.[13] This characterization needs further specification; there seem to be several different notions of necessity, e.g., logical necessity or natural necessity, metaphysical necessity or epistemic necessity. Specifying it using the notion of metaphysical necessity yields the following thesis:

---

12. The main point is that it treats the predicates "is a critic" and "is a member of" (or their paraphrases into elementary notation) as, so to speak, extensionally independent.

13. For example, Donald Kalish et. al. says "An argument is valid if it is impossible for its premisses to be true and its conclusion false" (1980, 1).

D. Sentence φ logically implies sentence ψ if, and only if, it is *metaphysically* necessary that if φ then ψ.

This thesis is relevant to Thesis *C*; on the standard view on the modal status of sets,

E. It is metaphysically necessary that if there is something x and if there is something y, then there is something (viz. $\{x, y\}$) of which x and y are members.[14]

Thesis *E*, together with standard principles of modal logic (which include the principles of elementary logic), yields the following:

It is metaphysically necessary that if Ezra is a critic and Thomas is a critic, there is something of which a critic is a member.

By Thesis *D*, "Ezra is a critic and Thomas is a critic" logically implies "There is something of which a critic is a member" (sentence (3)). If so, sentence (7) logically implies sentence (3), contrary to Thesis *C*, because (7) clearly logically implies "Ezra is a critic and Thomas is a critic."

This shows that Thesis *C* conflicts with Theses *D* and *E*, taken together. Now the objection holds that we must reject Thesis *C* while accepting Theses *D* and *E*. I disagree. It is one (or both) of these two theses that must be rejected. In particular, there is a good reason to reject Thesis *D*. The characterization of logical implication in terms of *metaphysical* necessity departs from a core idea underlying the standard account of metaphysical necessity.[15] On the account, metaphysical necessity is one thing, logical implication or validity quite another. In particular, it is one thing to say that a conditional states a metaphysical necessity, yet another to say that its antecedent logically implies its consequent.

To see this, consider some better-known cases. First, on the standard account, it is metaphysically necessary that if Cicero is Cicero, then Cicero is Tully. Nonetheless, "Cicero is Cicero" does not logically imply "Cicero is Tully"; that is, the following argument is not logically valid:

---

14. See Kit Fine (1981) for the standard view.
15. For the standard account, see, e.g., Saul Kripke (1980).

Cicero is Cicero.
Therefore, Cicero is Tully.

Similarly, "Chelsea is a critic" does not logically imply "Chelsea is not Hillary", although it is metaphysically necessary, on the same account, that if Chelsea is a critic, then Chelsea is not Hillary. Second, the standard account also holds that biological origin is a necessary character of an organism. That is, it is metaphysically necessary that, for example, if Chelsea, born by Hillary, exists, then Hillary is her biological mother. But this does not mean that the following arguments, for example, are valid:

Chelsea exists.
Therefore, Hillary is Chelsea's mother.

Chelsea lives in Washington.
Therefore, Hillary is a mother of something that lives in Washington.

Chelsea lives in Washington.
Therefore, there is something that is a mother of something that lives in Washington.

I maintain that the same separation between metaphysical necessity and logical correctness applies to the status of sets, e.g., the set {Ezra, Thomas}. It is one thing to hold that it is metaphysically necessary that

> If Ezra is a critic and Thomas is a critic, then there is something of which a critic is a member.[16]

---

16. Notice that Thesis *E* yields the following as further metaphysical necessities:

> If Ezra is a critic, there is something of which a critic is a member.
>
> If Ezra is Ezra, there is something of which something is a member.
>
> If Ezra is a critic, there is something (e.g., {Ezra, {Ezra}}) of which both a critic and something of which a critic is a member are members.

It is yet another to hold that the following argument is logically valid:

> Ezra is a critic and Thomas is a critic.
> Therefore, there is something of which a critic is a member.

This argument is no more valid than the other arguments mentioned above.[17]

1.3 Paraphrase

I now argue against Quine's paraphrase of the Geach-Kaplan sentence (1) using the result that this sentence does not logically imply "There is something of which a critic is a member." Recall the sentence in elementary notation that Quine takes to paraphrase sentence (1):

(2) $\exists a(\exists x x \varepsilon a \wedge \forall x[x\varepsilon a \rightarrow x \text{ is-a-critic} \wedge \forall y(x \text{ admires } y \rightarrow x \neq y \wedge y\varepsilon a)])$.

This sentence logically implies the following on elementary logic:

(4) $\exists a \exists x(x \text{ is-a-critic} \wedge x\varepsilon a)$.

And this sentence is a straightforward paraphrase of the following into elementary notation:

(3) There is something of which a critic is a member.

Thus a sentence that can be paraphrased by sentence (2) into elementary notation must logically imply sentence (3). But sentence (1) does not do so, as the Implication Argument shows. Consequently, it cannot be correctly paraphrased by sentence (2).

This argument extends the result of the Implication Argument with the help of an important principle regarding paraphrases of English sentences into elementary notation:

---

17. For discussions of various limitations of metaphysical modalities, see K. Fine (1994) and Joseph Almog (1991).

> *Preservation Principle*: Paraphrasing into elementary notation must preserve logical relations among paraphrased sentences.

The italicized "Thus" in the preceding paragraph indicates application of the principle. Now one might ask what is the reason for accepting the Preservation Principle. The answer is: the principle merely codifies what is taken for granted in important uses of paraphrases into elementary notation.

To test whether arguments formulated in natural languages are logically valid, it is usual to consider the arguments that paraphrase them into elementary notation, namely, the arguments whose premisses and conclusions are paraphrases of those of the original arguments into elementary notation. Then the results of applying elementary logic to the arguments formulated in elementary notation are transferred to the original arguments. For example, the first of the following arguments is paraphrased by the second into elementary notation:

> Ezra is a critic who admires only another critic.
> Therefore, there is a critic who admires only another critic.

> Ezra is-a-critic $\wedge$ $\forall x$[Ezra admires $x \rightarrow$ Ezra$\neq x \wedge x$ is-a-critic].
> $\therefore$ $\exists y$(y is-a-critic $\wedge$ $\forall x$[y admires $x \rightarrow y \neq x \wedge x$ is-a-critic]).

Thus we transfer the result of elementary logic, that the second argument is logically valid, and conclude that the first argument is also logically valid. To draw this conclusion, we must rely on the correctness of the Preservation Principle. We can see a pertinent example of the implicit reliance on the Preservation Principle in Quine's discussion of the Geach-Kaplan sentence:

> Kaplan has proved that we cannot express [the Geach-Kaplan sentence] in terms of just ['critic'], 'admire', identity, truth functions, and quantification. [The sentence] does not imply that there are two [critics] who admire only each other. It might hold because of some irreducible group of say eleven mutual admirers. (1982, 293)

In this passage, Quine explains why the Geach-Kaplan sentence cannot be paraphrased by a sentence used to paraphrase "There are two critics who admire only each other" into elementary notation. The reason that he gives is: "There are *some* critics who admire only one another" does not logically imply "There are *two* critics who admire only each other." Clearly, this explanation rests on the principle. There is then no objecting to the Preservation Principle. If so, it follows that the Geach-Kaplan sentence cannot be paraphrased by the singular quantification (2).

2. The Logic of Plurals

Elementary logic has conspicuous limitations with respect to plural constructions. It is helpless with, for example, the following arguments:

> Ezra and Thomas are critics who admire only each other.
> Therefore, some critics admire only one another.
>
> John and Carol are children who carry a piano.
> Therefore, some children carry a piano.

These arguments can be as easily seen to be logically valid as the valid arguments given proper treatments by elementary logic, such as the following:

> John is a child who is healthy.
> Therefore, a child is healthy.

In this chapter, I present a system of logic that, as a conservative extension of elementary logic, does equal justice to arguments that rely on plural constructions, such as the first two.

2.1 Language

The development of elementary logic is accompanied by that of the standard system of notation that I call *elementary notation*. In contrast to sentences in natural languages, such as English, sentences in elementary notation conform to a fixed number of austere syntactic structures, which helps to facilitate the analysis and formulation of the logical properties of, and relations among, the sentences. Thus it is usual to formulate elementary logic with respect to sentences in elementary notation while leaving its application to, e.g., English sentences to rest on their judicious paraphrases into the notation. The limitations here in question of elementary logic stem ultimately from the limited expressive power of elementary notation.

Elementary notation is designed primarily for singular constructions of natural languages, such as the premiss and conclusion of the third argument mentioned above:

(1) John is a child who is healthy.
(2) A child is healthy.

These sentences can be straightforwardly paraphrased into the notation:

(1') John is-a-child and John is-healthy (in symbols, $[C(j) \wedge H(j)]$);
(2') There-is-something x such-that [x is-a-child and x is-healthy] (in symbols, $\exists x[C(x) \wedge H(x)]$);

where "is-a-child" (in symbols, "C") and "is-healthy" (in symbols, "H") are simple predicates in the notation that correspond to the predicate phrases "is a child" and "is healthy" in English. By contrast, the notation does not have adequate paraphrases of such plural constructions as the premiss and conclusion of the second argument mentioned above:

(3) John and Carol are children who carry a piano.
(4) Some children carry a piano.[1]

1. Here I use sentences (3) and (4) to mean the following, respectively:

> John and Carol are children such that there is a piano (e.g., Bob) that they (i.e., John and Carol) *carry*.

> There are some children such that there is a piano (e.g., Bob) that they (i.e., the children) *carry*.

Consequently, elementary logic is not applicable to the argument.

In this section, I examine the limitations of elementary notation with respect to plural constructions and give a natural extension of the notation. I call the extended notation the *canonical notation for plurals*, because it is capable of natural paraphrases of a wide variety of plural constructions in English, such as (3) and (4). In the next section, I present the desired extension of elementary logic, which I call *the logic of plurals*. The logic is formulated with respect to sentences in the canonical notation (for plurals); the syntactic clarity of such sentences facilitates analysis as well as statements of their logical properties or relations. But the logic, by design, is applicable to a substantial fragment of English via judicious paraphrases just as elementary logic is so applicable to its smaller fragment.

2.1.1 Elementary Notation

Elementary notation, which elementary logic is usually couched in, can be taken to contain basic resources of five kinds:

(a) *Singular Constants*
"John" (or "*j*"), "Carol" (or "*c*"), etc.

(b) *Singular Variables*
"*x*", "*y*", etc.

(c) *Predicates*

i. *1-place predicates*
"is-a-child" (or "*C*"), "is-healthy" (or "*H*"), etc.

ii. *2-place predicates*
"is-identical-with" (or "="), "carries" (or "C_1"), "works-with" (or "*W*"), etc.

iii. *3-place predicates*
"gives . . . to" (or "*G*"), etc.

. . . .

(d) *Truth-functional Sentential Connectives*

And throughout this chapter, I use "carry" in the sense that "John and Carol carry Bob" is true if John and Carry collaborate and carry Bob together while neither of them alone carries it.

The conjunction symbol "and" (or "∧"), the negation symbol "It-is-not-the-case-that" (or "~"), etc.

(e) *Standard Quantifiers*
The standard existential quantifier "There-is-something . . . such-that" (or "∃")
The standard universal quantifier "Anything . . . is-such-that" (or "∀")

Sentences in the notation are constructed from these resources in the familiar way.[2] A simple sentence is formed by an *n*-place predicate and *n*-many occurrences of singular constants or variables, such as "*x* is-a-child" (or "$C(x)$") or "John works-with *y*" (or "$W(j, y)$"). A standard quantifier (e.g., "There-is-something . . . such-that"), a variable (e.g., "*x*"), and a sentence in the notation (e.g., "$C(x)$") form a complex sentence, such as the existential generalization "*There-is-something x such-that x* is-a-child" (or "$\exists xC(x)$"). Finally, a sentential connective combines with an appropriate number of sentences to form a complex sentence, a negation, such as "*It-is-not-the-case-that x* is-a-child" (or "$\sim C(x)$"), or a conjunction, such as "*x* is-a-child *and* John works-with *y*" (or "$[C(x) \wedge W(j, y)]$").[3]

2.1.2 The Term Connective "and"

Notice that elementary notation has no counterpart of the connective "and" used in, e.g., the following sentence:

(3) John *and* Carol are children who carry a piano.

---

2. Here I use "sentence (in elementary notation)" in a broader sense. On the usual terminology, "*x* is-a-child" (or "$C(x)$")", for example, is not a sentence (i.e., a closed formula) but only a formula; on my terminology, it is a sentence but not a closed one. I avoid the word "formula" because it is usual to associate it with the idea that a formula (unlike English sentences) is a sequence of *meaningless* 'symbols'.

3. I ignore parentheses in the discussion of syntax. They can be considered parts of sentential connectives by identifying the conjunction symbol, for example, as "[. . . ∧ . . .]".

In this sentence, "and" works on two singular terms, "John" and "Carol", and yields a complex term, "John and Carol". Call such a connective, which operates on two or more terms to yield a complex term, a *term connective*. This use of "and" contrasts with its use in the following:

(5) John is a child *and* Carol is a child.

In this sentence, "and" works as a *sentential connective*; it 'connects' two sentences to yield a more complex sentence. Now, elementary notation allows only the connective "and" as used as a sentential connective. I extend the notation by adding a term connective "**and**" (in symbols, "@") that corresponds to the use of "and" in English as a term connective. The connective yields plural terms, such as "John **and** Carol" (in symbols, "[j@c]"), in the canonical notation.

Let me explain why the sentential connective alone would not do. The lack of the term connective in elementary notation is compensated to some extent by a close interplay between the two uses of "and" in English. Consider the following sentence:

(6) John *and* Carol are children.

Although "and" is used here as a term connective, it is usual to paraphrase the sentence into elementary notation as follows:

(5') John is-a-child *and* Carol is-a-child (in symbols, $[C(j)\wedge C(c)]$),

which is the straightforward paraphrase of the sentential conjunction "John is a child *and* Carol is a child" (sentence (5)). The basis for this paraphrase of (6) lies in the connection between the two English sentences equally paraphrased by (5'): because the plural construction (6) can be rendered to the singular construction (5), the paraphrase of (5) into elementary notation can be taken to paraphrase (6) as well. But the pleasing interplay between the sentential connective and the term connective uses of "and" does not hold in general. Consider, for example, the following sentence, where "Bob" is the name of a huge piano:

(7) John *and* Carol carry Bob.

This sentence cannot be rephrased as follows:

(8) John carries Bob *and* Carol carries Bob.

For (7) would be true while (8) being false, if John and Carol got together and cooperated to carry Bob, which neither of the children can carry alone. Thus sentence (7) cannot be paraphrased into elementary notation in the way that (6) is so paraphrased; the usual paraphrase of sentence (8) into elementary notation cannot be accepted as a paraphrase of (7).

Surely, the attempt to paraphrase "John and Carol are children and John and Carol carry a piano" (sentence (3)) meets the same difficulty. The task of paraphrasing it using only the sentential connective reduces to that of paraphrasing away the italicized occurrence of "and" in "John and Carol are children and there is a piano that John *and* Carol carry." But that occurrence of "and" cannot be eliminated in favor of "and" used as a sentential connective.[4]

2.1.3 Predicates

The canonical notation for plurals has a predicate that corresponds to the English predicate that occurs in "John and Carol carry Bob" (sentence (7)): "to **carry**" (in symbols, "**R**"). Thus the sentence can be paraphrased into the notation as follows:

(7') [John **and** Carol] **carry** Bob (in symbols, $\mathbf{R}([j@c], b)$).

I distinguish the predicate "to **carry**" that is used to paraphrase sentence (7) into the canonical notation from the predicate "carries" (or "C_1") in the usual paraphrase of "John carries Bob" into elementary notation. The two predicates, though related, differ in grammatical character. To explain what kinds of predicates are available in the canonical notation, it is necessary to clarify their grammatical difference.

---

4. "There is a piano that John and Carol carry" contrasts with, e.g., "There is a man whom John *and* Carol are friends of", which can be paraphrased by "There is a man such that John is a friend of his and that Carol is a friend of his."

2.1.3.1 Singular, Plural, and Neutral Predicates

The grammatical distinction can be found among predicates (or else their various forms) in, e.g., English. Let me first illustrate and draw the distinctions with respect to predicates in English. Then it is straightforward to carry them over to predicates in the canonical notation.

Compare "John and Carol *carry* Bob" (sentence (7)) with the following sentences:

> John, Carol, and Bill *carry* Bob.
> John, Carol, Bill, and Hillary *carry* Bob.
> John, Carol, Bill, Hillary, and Chelsea *carry* Bob.

It should be clear that these sentences have the same predicate as the predicate of sentence (7).[5] If so, the common predicate "carry" (or "to carry") must be a *2-place predicate*, with two argument places to be filled in to form a grammatical sentence. In the above-mentioned sentences, including (7), its second argument place is filled in by the same singular term "Bob" and its first argument place by "John and Carol", "John, Carol, and Bill", and so on, respectively. Now, consider another sentence:

> (9) John *carries* Bob.

Does this sentence, too, have the same predicate? There is, it seems, room for disagreement over this question, disagreement on how to individuate predicates (or, in general, words) in English. Some people might hold that the predicate of the singular construction (9) is literally the same as that of the plural construction (7): "to carry". On this view, the common predicate can take two different *forms*, (present) singular and plural, depending on the grammatical number of the term filling in its first argument place. Some might disagree; they might think that the so-called forms of the putative predicate "to carry" are actually two different, if related, predicates. On their view, the predicates of sentences (9) and (7) are "carries" and "carry", respectively.

---

5. It is this grammatical fact that yields the ultimate justification of the view that "and" is used in sentence (7) as a term connective that yields the complex term "John and Carol" as a grammatical unit of the sentence.

I am inclined to accept the first view, but my present aim is not to establish that this is the correct view on the nature of English predicates as they stand. My aim is to use the three putative predicates entertained by those who hold one or the other of the two views to illustrate an important distinction among predicates and their argument places.

Consider the grammatical character of the first argument places of the three putative predicates: "to carry", "carries", and "carry". First, the (putative) predicate "carrie*s*" admits only a singular term into its first argument place. That is, "John and Carol *carries* (*not*: carry) Bob" is not a grammatical sentence. Call such an argument place of a predicate *singular*. Second, it is the other way around with the first argument place of "carry". It does not fit in with a singular term; "John *carry* Bob", for example, is not grammatical. Call such an argument place *plural*. Finally, the (putative) predicate "to carry" admits either a singular or a plural term into its first argument place. The predicate takes the plural form if the argument place is filled in with a plural term, such as "John and Carol", and the singular form to combine with a singular term, such as "John". Call such an argument place *neutral*.[6] And we can classify predicates on the basis of what kind of argument places they have. Say that a predicate is *singular* (*plural* or *neutral*) with respect to an argument place, if the argument place is *singular* (*plural* or *neutral*). Accordingly, say that a predicate is singular (*plural* or *neutral*), if all of its argument places are singular (*plural* or *neutral*).

Now it is straightforward to carry this distinction over to the canonical notation. The notation has plural terms (e.g., "John **and** Carol") as well as singular terms (e.g., "John") and draws a syntactic distinction between them. So the above distinctions on predicates and their argument places are applicable to the predicates in the canonical notation and their argument places as well.

I can now clarify the remark made above on the predicates "to **carry**" (used to paraphrase sentence (7): "John and Carol carry Bob") and "carries" (used in the usual paraphrase of sentence (9): "John carries Bob"). The predicate "to **carry**" is introduced into the canonical notation as a *neutral* predicate; it is the counterpart of the predicate "to carry" that occurs in both the plural and singular constructions (7) and (9) on the first view discussed above. By contrast, the predicate "carries" (or "C_1 ") available in elementary notation is identified as a *singular* predicate in the canonical notation. The ground for this identification lies in the idea that the predicate is introduced into elementary notation as a

6. A better example of a neutral argument place would be the second one, placed after "by", of "to be carried by".

counterpart of the singular form "carries" of the (putative) predicate "to carry". It is the same with the other predicates in elementary notation; they can all be identified as singular predicates in the canonical notation.

Notice also that the neutral predicate "to **carry**", instead of a plural predicate, is used in paraphrasing plural constructions, such as sentence (7), into the canonical notation. This, too, illustrates the general scheme that I adopt in paraphrasing English predicates. I design the canonical notation without predicates with plural argument places, and paraphrase such predicates in English, if it has any such predicates, into the canonical notation using the neutral predicates whose corresponding argument places are neutral.[7]

2.1.3.2 Neutral Expansions and Singular Reducts

Recall sentence (6), "John and Carol are children." I regard this sentence to contain the complex term "John and Carol" that occurs in sentence (7), "John and Carol carry Bob."

This view contrasts with the view implicit in the usual paraphrase of sentence (6) into elementary notation. What I have in mind is the view that the term connective in (6) is merely a device for an abbreviated expression of the sentential connective in "John is a child *and* Carol is a child" (sentence (5)). One cannot maintain this view with regard to the term connective "and" that occurs in sentence (7), but one might still hold that "and" is used as an abbreviation device in sentence (6) by distinguishing two different functions (or sub-uses) of the term connective use of "and". This bifurcation of the term connective use of "and" finds its ground in a handy account of the limited interplay between the sentential and term connective uses of "and". Sentence (5), containing the sentential connective, can be given as a mere elaboration of sentence (6), in which "and" occurs as a term connective. One might argue that this is because (6) is merely an abbreviated or contracted form of (5).

But the bifurcation view encounters serious difficulties in doing justice to other equally legitimate contractions. Consider the following sentences:

---

7. On the second view mentioned above, the plural form "carry" is a plural predicate ("to play together", taken as a single predicate, seems to be a better example of such a predicate). Thus the canonical notation might seem to deviate from English in this respect, but there is no difficulty in extending the notation to include predicates with plural argument places as well.

(10) John and_1 Carol are children and John and_2 Carol carry Bob.
(11) John and_3 Carol are children who carry Bob.

These two sentences contract or expand to each other just as well as sentences (5) and (6) do so. But one cannot accept (11) as resulting from a legitimate abbreviation of (10) while also maintaining that the plural terms "John and_1 Carol" and "John and_2 Carol" are of different kinds on account of the difference between the homonymous term connectives "and_1" and "and_2". Doing so is as objectionable as taking the plural construction "John and Carol go to $banks_3$" to derive from the heterogeneous conjunction "John goes to a $bank_1$ and Carol goes to a $bank_2$", where "$bank_1$" and "$bank_2$" are used ambiguously for money-banks and river-banks, respectively.

Thus I reject the bifurcation view of the term connective and the abbreviation explanation that leads to it. If so, how can I explain the connection between (5) and (6)? I explain it by considering the nature of their predicates.

Here is the phenomenon that requires explanation: "John and Carol *are children*" (sentence (6)) can be paraphrased by "John *is a child* and Carol *is a child*" (sentence (5)); but "John and Carol *carry Bob*" (sentence (7)) cannot be paraphrased by "John *carries Bob* and Carol *carries Bob*" (sentence (8)). Notice, first, that we can give an alternative description of this phenomenon by considering different behaviors of the two (putative) predicates: "to be a child or children" (in short, "to be a-child/children" or "to be-child[ren]"), taken to be common to (5) and (6), and "to carry", taken to be common to (7) and (8). The first predicate *distributes over* the term connective "and" with respect to its first argument place (that is, (5) and (6), for example, are logically equivalent),[8] but it is not so with the second predicate. Now, I explain the distributivity of "to be-child[ren]" as due to its special character. This predicate must be taken to derive from the underlying singular predicate "is a child" in a special way. It is a neutral predicate designed to be true of some (one or more) things (taken together) if and only if the singular predicate "is a child" is true of each one of them. So the neutral predicate is true of, e.g., John *and* Carol (taken together) if and only if "is a child" is true of John *and also of* Carol. Hence, its distributivity. By contrast, the predicate "to carry", which is also neutral with respect to its first argument place, is not related to its singular counterpart "carries" in a similar

8. Compare this with: the multiplication operation distributing over the addition operation so that, for example, 2x(3+5)=2x3+2x5.

way. It is convenient to put this point in terms of the logical behavior of the related complex predicates "to carry Bob" and "carries Bob". These two predicates are not related in such a way that the former is true of some (one or more) things (taken together) if and only if the latter is true of each one of them. So the neutral predicate "to carry" cannot be seen to result from the singular predicate "carries" (with respect to its first argument place) as "to be-child[ren]" results from "is a child".[9] In this case, I suggest, the neutral predicate "to carry" must be taken to be basic and the singular predicate "carries" must be taken to be derivative. To put it figuratively, the singular predicate is the 'projection' of the neutral one onto the horizon of singular terms. Such neutral predicates do not in general distribute over the term connective "and".

This completes my explanation of the different logical behaviors of "to be-child[ren]" and "to carry". The explanation relies on postulating the existence of neutral predicates of a special kind, such as "to be-child[ren]". Call such predicates the *neutral expansions* of the underlying predicates (with respect to the relevant argument places). The explanation also leads to a distinct kind of singular predicates, such as "carries". Call such predicates the *singular reducts* of the underlying predicates (with respect to the relevant argument place).[10]

---

9. By considering the complex one-place predicates, I avoid complications arising from such examples as "John and Carol carry pianos." This sentence (used in one way) is logically equivalent with "John carries a piano (or pianos) and Carol carries a piano (or pianos)." But this does not conflict with the view that the predicate "to carry" that occurs in sentence (7), "John and Carol carry Bob", as I use the sentence (see footnote 1), fails to distribute over the term connective "and" with respect to its first argument place. Even the sentence (7) as used in a different way (in which I believe it is rare to use the sentence) is logically equivalent with "John carries Bob and Carol carries Bob." The sentence (7) as used in this way, I think, contains a predicate only homonymous with the one contained in the sentence used in the way I have in mind. Using the terminology introduced in the next paragraph, I consider the former predicate as the neutral expansion of the singular reduct of the latter predicate. Similarly, "John can Carol carry pianos" (used in the way in question) seems to contain the neutral expansion of the complex singular predicate "carries a-piano/pianos".

10. Here I leave it open whether there are distributive neutral predicates that are not the neutral expansions. I think that there are such predicates (consider, for example, "to be finitely many human beings"), but they are more complex than those I need to discuss in this work.

Now, these distinctions have logical significance. So the canonical notation incorporates the distinctions. The notation has a way of designating neutral expansions. For example, the predicate in the notation corresponding to the English predicate "to be-child[ren]" is "is-a-child^{N}" (or "C^{N}")", where the superscript "N" indicates that the predicate is related to the singular predicate "is-a-child", available in elementary notation, as its neutral expansion. It has a way of designating singular reducts as well. For example, the singular reduct of the neutral predicate "to **carry**" (or "**R**") in the canonical notation can be written as "to **carry**^{S}" (or "$\mathbf{R}^{S}$"), which can be identified with the predicate "carries" (or "C_1") available in elementary notation. Thus sentences (6) and (7) can be paraphrased into the notation as follows:

(6') [John **and** Carol] is-a-child^{N} (in symbols, $C^{N}([j@c])$).
(7') [John **and** Carol] **carry** Bob (in symbols, $\mathbf{R}([j@c], b)$).

The paraphrases reflect the view that the sentences contain the same plural term "John and Carol" but that their predicates differ with respect to the argument places that the term fills in.

2.1.3.3 Logical Predicates

Elementary notation has one predicate with special logical significance: the predicate for identity "is-identical-with" (in symbols, "="). Elementary logic treats this predicate as a logical one to account for the validity of some arguments as relying on its logicality. The extended logic that I shall present gives the same status to two predicates that are available only in the canonical notation: "to **be-the-same-as**" (in symbols, "**S**") and "**is-one-of**" (in symbols, "$\boldsymbol{\eta}$").[11]

The singular two-place predicate "is-identical-with" (or "=") is used to paraphrase "John *is* not Carol" or "Cicero *is* Tully", for example, into elementary notation. Now, consider the following sentences: "John and Bob *are* not Bob and Carol", "Cicero and John *are* John and Tully", and "They *are* John

11. There is another logical predicate, the counterpart of "to be among". It can be considered the neutral expansion of "**is-one-of**" (with respect to its first argument place).

and Tully."[12] These sentences are the plural counterparts of the former sentences: they have the plural form "are" of the verb "to be" (flanked by plural terms), whereas the former sentences have its singular form "is" (flanked by singular terms). The canonical notation contains the quasi-identity predicate "to **be-the-same-as**" (or "**S**") as the neutral counterpart of the singular predicate for identity. So "Cicero and John are John and Tully", for example, can be paraphrased into the canonical notation as follows:

> [Cicero **and** John] **be-the-same-as** [John **and** Tully] (in symbols, $\mathbf{S}([c_1@j], [j@t])$)

where "c_1", "j", and "t" stand in for "Cicero", "John", and "Tully", respectively.[13]

The predicate "**is-one-of**" (or "η"), whose second argument place is neutral while its first argument place is singular, is incorporated into the canonical notation to paraphrase such sentences as "John and Tully are healthy and *one of* them is Cicero" or "There are some Americans who are parents of Chelsea and

---

12. The following are sentences used more naturally that are to be paraphrased in terms of "to **be-the-same-as**": "The parents of Chelsea *are* Bill and Hillary" and "The children in the Common Room *are* John and Carol." Their singular counterparts, usually paraphrased in terms of "is-identical-with", are: "The mother of Chelsea *is* Hillary" and "The child in the Common Room *is* John." I avoid giving these sentences as examples in the text because elementary notation and the canonical notation do not have direct counterparts of definite descriptions, singular or plural.

13. Notice that the usual distinction between the two uses of "is" applies to the plural form "are" as well. The singular form "is" is used as the predicate for identity in "Cicero *is* Tully" but merely as an indicator of predication in "John *is* healthy" or "Cicero *is* not alive." Similarly, the plural form "are" is used in the sentences listed above in a way different from the way it is used in, e.g., "John and Carol *are* children" or "They *are* healthy." Just as the predicate "is-identical-with" corresponds only to the first use of "is", so does the neutral predicate "to **be-the-same-as**" amount only to the first use of the plural form "are" (or, more precisely, the predicate "to be").

Bill *is one of* them."[14] The former sentence, for example, can be paraphrased into the canonical notation as follows:

> [John **and** Tully] is-healthy^{N} $\wedge$ Cicero **is-one-of** [John **and** Tully] (in symbols, $H^{N}([j@t]) \wedge \eta(c_1, [j@t])$).

Notice that the predicate "to **be-the-same-as**" merits the distinguished status on account of its clear connection to the predicate for identity. The predicate "**is-one-of**", too, turns out to have a close connection to the identity predicate by way of its relation to the quasi-identity predicate. For example, John and Carol *are* not Bob and Carol, because *one of* the former (viz. John) is not *one of* the latter; and conversely, Cicero and Carol *are* Carol and Tully, because each one of the former two is also one of the latter two (and *vice versa*).[15]

2.1.4 Quantifiers and Variables

Another extension of elementary notation is needed to treat quantified sentences, such as the following:

> (12) *Some* children carry Bob.

The word "some" used in this sentence is in a way the plural form of the word "a" used in, e.g., the following:

> (2) *A* child is healthy.

Sentence (2) is paraphrased into elementary notation in terms of the standard existential quantifier and an associated variable. The canonical notation contains

---

14. One must clearly distinguish this predicate from the predicate "is-a-member-of" (in symbols, "ε") that indicates the membership relation that a member (e.g., John) of a set (e.g., {John}) has to the set. This is a singular predicate available in elementary notation.

15. Thus we can just add the predicate "**is-one-of**" and use it to define "to **be-the-same-as**".

the plural counterparts of these resources, and they help to yield a paraphrase of sentence (12) in a similar way.

2.1.4.1 Singular and Plural Quantifiers

Both "some" in (12) and "a" in (2) are existential quantifiers, which contrast with the universal quantifier "every" used in, for example, "Every child is healthy." But "some" in (12) goes with plural constructions, such as nouns or predicates in the plural form (e.g., "children" or "carry"), whereas "a" goes with singular constructions (e.g., "child" or "is healthy"). Call quantifiers of the two kinds *plural* and *singular*, respectively. Accordingly, call such sentences as (12) and (2) *plural* and *singular quantifications*, respectively. Notice that this distinction between singular and plural quantifiers does not directly apply to the quantifiers in elementary notation. The notation, strictly speaking, does not draw a distinction between singular and plural forms of nouns or predicates. Still, as my practice indicates, it is reasonable to identify the quantifiers as singular quantifiers and associate them with singular quantifier phrases in English (e.g., "There is something . . . such that") instead of their plural counterparts (e.g., "There are some things . . . such that"). The reason for this is that the variables in elementary notation, which fill in their first argument places, are refinements of *singular pronouns*, such as "he", "she", "it", "this", or "that". Let me explain.

Consider pronouns used for cross-reference in, for example, "There is a child and *it* is healthy" or "There is a cat and there is a dog over there and he is chasing her." We might take a scheme of clarifying which phrases are the antecedents of the pronouns. One such way, though one not practiced in English, is to tag the pronouns to their antecedents as follows: "There is a cat [*she*] and there is a dog [*he*] over there and *he* is chasing *her*", where the pronouns "she" and "he" that occur in the parentheses are tagged to "a cat" and "a dog", respectively, so that these phrases are the antecedents of their later occurrences. Variables in elementary notation are essentially pronouns used in this way. Recall how variables combine with quantifiers in elementary notation, such as "There-is-something . . . such-that" (or "∃"). The quantifiers have two argument places. The first argument place, indicated by " . . .", is to be filled in by a variable in the notation; the second by a sentence in the notation. In the usual paraphrase "There-is-something x such-that [x is-a-child and x is-healthy]" (in symbols, "$\exists x[C(x) \wedge H(x)]$") of sentence (2), for example, the variable "x" fills in the first argument place while the conjunction "x is-a-child and x is-healthy" does the second. What I suggest is to see the first argument place as the slot for tagging a pronoun to make it clear that the quantifier is the antecedent of its later occurrences.

Now, notice that English, for example, has plural pronouns, such as "they", "these" and "those", as well as singular pronouns. And the use of pronouns described above is not limited to the singular ones. The plural pronoun "they" is used in the same way in, for example, "There are *some children* and *they* carry Bob." Still, the variables in elementary notation turn out to be refinements of only the singular pronouns so used and that the quantifiers in elementary notation are tagged with those variables. So the quantifiers can be taken to amount to singular quantifiers, which can be the antecedents of singular pronouns.

This understanding suggests the following criterion. Call a variable *singular*, if it is a counterpart of a singular pronoun (used as explained above); and *plural*, if it is a counterpart of a plural pronoun. Accordingly, call a quantifier whose first argument place is to be filled in by a variable *singular* if the argument place admits only singular variables; and *plural* if the argument place admits only plural variables. This criterion yields the desired result that the quantifiers, as well as the variables, in elementary notation are singular.[16]

2.1.4.2 Paraphrasing Plural Quantifiers

In addition to the singular variables and quantifiers in elementary notation, the canonical notation has their plural counterparts:

(i) plural variables: "*xs*", "*ys*", etc.
(ii) plural quantifiers: the plural existential quantifier "There-are-some-things . . . such-that" (or "Σ") and the plural universal quantifier "Any-things . . . are-such-that" (or "Π").

---

16. One might conceive another possibility: *neutral* quantifiers that admit both singular and plural variables. Such quantifiers can be regarded as taking singular or plural *forms* depending on whether its argument place is filled in with singular or plural variables. Now, the standard quantifiers may be taken as neutral quantifiers, though they always take singular forms in sentences in elementary notation. I ignore this third possibility in the text, but I regard the decision as a matter of convenience in exposition.

They are used to paraphrase plural quantifications of, e.g., English. For example, "Some children carry Bob" (sentence (12)) can be paraphrased into the notation as follows:

(12') There-are-some-things *xs* such-that [*xs* is-a-child^{N} and *xs* **carry** Bob] (in symbols, $\Sigma xs[C^{N}(xs) \wedge \mathbf{R}(xs, b)]$).

This paraphrase of the plural quantification can be obtained just as the usual paraphrases of singular quantifications into elementary notation are obtained. And the ground for accepting (12') as a paraphrase of (12) is of the same kind as the one for accepting the usual paraphrases of singular quantifications into elementary notation. Let me explain.

Consider the usual paraphrase of, for example, "A child is healthy" (sentence (2)):

(2') There-is-something *x* such-that [*x* is-a-child $\wedge$ *x* is-healthy].

We can conceive a sequence of sentences leading from sentence (2) to sentence (2'):

(a) A child is healthy. (sentence (2))
(b) There is a child who is healthy.
(c) There is a child and he is healthy.
(d) There is something that is a child and it is healthy.
(e) There is something such that it is a child and that it is healthy.
(f) There is something *x* such that [*x* is a child and *x* is healthy].
(g) There-is-something *x* such-that [*x* is-a-child $\wedge$ *x* is-healthy]. (sentence (2'))

Any two consecutive sentences in the sequence can be seen to abbreviate or elaborate each other, either in English as it stands or else given clarifications of the nature of the quasi-English devices (e.g., the variable "*x*" in (f)) used in the later sentences in connection with the corresponding devices (e.g., the pronoun "it" in (e)) that are used in the earlier sentences. Thus the clarifications of the new devices and connections already found in English provide the ultimate justification for paraphrasing (2) by (2').

It should be clear that the same justification can be given for paraphrasing the English plural quantification (12), for example, by (12'). The only essential difference comes in justifying two steps: (i) replacing the plural pronoun "they" with the plural variable "*xs*" and (ii) replacing the English quantifier phrase "There are some things . . . such that" with the plural quantifier "There-are-some-things . . . such-that" in the canonical notation. The requisite justifications come from clarifications of the nature of the two devices in connection with their English counterparts. The plural quantifiers and variables, like their singular counterparts, are introduced in tandem into the canonical notation as refinements and standardizations of certain English quantifier phrases and the English pronouns that take them as antecedents.

2.1.4.3 Singularizable Plural Quantifications

Consider another plural quantification in English:

(13) *Some* children are healthy.

It is usual to paraphrase it into elementary notation as follows:

(2') There-is-something x such-that [x is-a-child and x is-healthy] (in symbols, $\exists x[C(x) \wedge H(x)]$).

But the canonical notation has a straightforward paraphrase of (13):

(13') There-are-some-things *xs* such-that [*xs* is-a-child^{N} and *xs* is-healthy^{N}] (in symbols, $\Sigma xs[C^{N}(xs) \wedge H^{N}(xs)]$).

Natural elaborations of the English plural quantification (13), such as "There are some child*ren* and *they* are healthy", directly lead to the plural quantification (13') in the canonical notation. If so, what can be the justification for taking the singular quantification (2') to paraphrase (13)?

To take (2') to paraphrase (13), it is necessary to appeal to the logical connection (or a stronger relation) between (13) and its singular counterpart "A child is healthy" (sentence (2)). Natural elaborations of this sentence lead to the singular quantification (2') in elementary notation (for the variable "x" in (2') is a refinement of the pronoun "it" that occurs in natural elaborations of (2), such as

"There is something that is a child and it is healthy"), and the singular and plural quantifications (2) and (13) are logically equivalent in the same intuitive way that, for example, the sentential conjunction "John is a child *and* Carol is a child" and its plural cousin "John *and* Carol are children" are so. So, one might argue, we can use the paraphrase of (2) to paraphrase (13) as well.

Notice that this justification for taking (2') to paraphrase (13) must presuppose, and so cannot help to explain, the logical equivalence between (2) and (13). Thus we cannot use elementary logic to account for their logical equivalence. By contrast, we can use the logic of plurals, which is presented in the next section, to show that the natural paraphrases of (2) and (13), for example, into the canonical notation (i.e., (2') and (13')) are logically equivalent. The logic systematizes, and helps to explain, the intuitions behind the usual paraphrases of singularizable plural quantifications, such as (13).

2.1.5 The Canonical Notation for the Logic of Plurals

We have seen that the terms, predicates, and quantifiers in elementary notation can be identified with singular terms, predicates of the singular form, and singular quantifiers of, e.g., English, respectively. The canonical notation adds their plural or neutral counterparts and yields natural paraphrases of plural, as well as singular, constructions of English. In the next section, I shall formulate a conservative extension of elementary logic with respect to the canonical notation and show that it gives proper treatments of arguments relying on plural constructions of English. Before I undertake the task, it is necessary to give a systematic account of the primitive and complex expressions in the notation. The primary aim here is not to give further clarifications of how well-formed expressions in the notation relate to the English constructions that they are used to paraphrase, but to give a survey of the kind of primitive resources available in the notation and clarify the rules governing the construction of complex expressions, especially sentences, in it. To do so, it is convenient to rely more on symbolic expressions (e.g., "∃") than on the quasi-English expressions that they abbreviate (e.g., "There-is-something . . . such-that").

2.1.5.1 Terms

The canonical notation contains primitive singular terms available in elementary notation: *singular constants* and *singular variables*. In addition, it has plural terms of two kinds.

First, it has plural counterparts of singular variables: *plural variables*, such as "*xs*" or "*ys*". As refinements of plural pronouns in English (e.g., "they"), the plural variables are not complex but simple expressions. But I use the ending "s" to get the variables to suggest their grammatical affinity to, as well as differences from, the singular variables.

Second, the canonical notation has *complex terms*, of the form $\ulcorner[\tau@\mu]\urcorner$, obtained by flanking the term connective "@" (or "**and**") with any two terms, τ and μ, in the notation: "[John @ Carol]", "[John @ John]", "[John @ *x*]", "[John @ *xs*]", "[*xs* @ *ys*]", "[*x* @ [John @ Carol]]", and so on. Such terms are all *plural* terms in the notation.

2.1.5.2 Predicates and Simple Sentences

Predicates in the canonical notation have some general characters. First, a predicate in the notation has a fixed, finite number (greater than zero) of argument places. Second, each argument place of a predicate must be either singular or neutral.

Predicates in the canonical notation can be classified on the basis of their *arities*, i.e., the number of their argument places: 1-place predicates, 2-place predicates, and so on. Then further distinctions can be made on the basis of the kind of argument places that they have: a predicate is *singular* (or *neutral*) with respect to an argument place, if the argument place is *singular* (or *neutral*); it is *singular* (or *neutral*), if all of its argument places are singular (or *neutral*). The predicates in elementary notation are identified as singular predicates in the canonical notation.

The canonical notation draws some further distinctions among its predicates. Some predicates neutral with respect to an argument place are designated as *neutral expansions* of a predicate: if π is a predicate, then let $\pi^{N(i)}$ (or $\pi^{N(j(1), j(2), \ldots, j(k))}$) be the neutral expansion of π with respect to its *i*-th argument place (or the j(1)-th, j(2)-th, . . . , and j(*k*)-th argument places) and π^{N} the neutral expansion of π with respect to all of its argument places. Similarly, some predicates singular with respect to some of their argument places are designated as *singular reducts*: if π is a predicate neutral with respect to its i-th argument place, then let $\pi^{S(i)}$ (or $\pi^{S(j(1), j(2), \ldots, j(k))}$) be the singular reduct of π with respect to the i-th argument place (or the j(1)-th, j(2)-th, . . . , and j(k)-th argument places) and π^{S} the singular reduct of π with respect to all of its argument places.

Finally, some predicates are distinguished as logical predicates. In addition to the two-place singular predicate for identity "is-identical-with" (or "="), there are two such predicates: the predicate for quasi-identity "**S**" (or "to **be-the-same -as**") and "**η**" (or "**is-one-of**").

Now, *simple sentences* in the canonical notation are constructed from its terms and predicates. Say that a term τ is *appropriate for* an argument place of a predicate, if either of the following conditions hold:

(i) τ is a singular term.
(ii) τ is a plural term, and the argument place is neutral.

Then if π is an n-place predicate and $\tau_1, \tau_2, \ldots, \tau_n$ are terms appropriate for the respective (i.e., the 1st, the 2nd, . . . , and the n-th) argument places of π, then $\ulcorner \pi(\tau_1, \tau_2, \ldots, \tau_n) \urcorner$ is a simple sentence in the canonical notation.

2.1.5.3 Quantifiers and Complex Sentences

Construction of complex sentences in the canonical notation involves either sentential connectives or quantifiers.

All the sentential connectives in the canonical notation are available in elementary notation. And the rules of construction of complex sentences in the canonical notation are the same as those in elementary notation.

The canonical notation has two sorts of quantifiers. First, the singular existential and universal quantifiers "∃" and "∀"; second, the plural existential and universal quantifiers "Σ" and "Π". The singular quantifiers are available in elementary notation; the plural ones are not. Accordingly, it is necessary to make a modification to the rule governing the construction of quantifications. The necessary modification concerns what kind of variables fit in with a quantifier. Say that a variable τ *is appropriate for* a quantifier Q, if either of the following holds:

(i) τ is a singular variable, and Q is a singular quantifier.
(ii) τ is a plural variable, and Q is a plural quantifier.

Then the rule of construction of quantifications in the canonical notation can be stated as follows:

> If Q is a canonical quantifier, τ a variable appropriate for quantifier Q, and φ a sentence in the canonical notation, then $\ulcorner Q\tau\varphi \urcorner$ is a (complex) sentence in the canonical notation.

2.2 Logic

In this section, I formulate the logic of plurals with respect to the canonical notation for plurals. The logic is a conservative extension of elementary logic as formulated with respect to elementary notation. If a sentence in elementary notation is a *logical truth* on elementary logic, it is a logical truth (in the canonical notation) on the logic of plurals; likewise, if an argument formulated in elementary notation is (*logically*) *valid* on elementary logic, it is so on the logic of plurals, too. Further, if a sentence (or argument) in elementary notation fails to be a logical truth (or valid argument) on elementary logic, it fails to be a logical truth (or a valid argument) on the logic of plurals, too. Thus the logic of plurals differs from elementary logic with regard to only sentences (or arguments) that belong properly to the canonical notation. I shall give a partial axiomatization of the logic and give a model-theoretic characterization of it.

2.2.1 The Logic of Plurals: Partial Axiomatization

I present four axiomatic systems couched in the canonical notation: Systems A, B, C, and D. System A is the weakest system, from which I build up the other systems by adding more axioms successively to get the strongest system D. System D is designed to give a *good approximation* to the logic of plurals. Closed sentences that are theorems of the system are logical truths on the logic of plurals. But the system leaves out some logical truths. This is due to the nature of the logic of plurals: No system of the kind here in question can give a complete characterization of the logic, although a complete characterization can be given in model-theoretic terms. In this respect, the logic of plurals, like higher-order logics, differs from elementary logic, which can be given a complete axiomatization.

2.2.1.1 System A

Axioms of System A are sentences of three groups in the canonical notation. Those of the first group constitute the *logic of truth functions*:

I. Truth-functional tautologies.

The axioms of the second group correspond to *the logic of singular quantifications*. They are instances of four schemata:[17]

II. $\varphi(\varsigma) \rightarrow \exists v \varphi(v)$, where ς is substitutable for v in $\varphi(v)$.
III. $\forall v[\varphi(v) \rightarrow \psi(v)] \rightarrow [\forall v \varphi(v) \rightarrow \forall v \psi(v)]$.
IV. $\varphi \rightarrow \forall v \varphi$, where v does not occur free in φ.
V. $\forall v \varphi(v) \leftrightarrow \sim\exists v \sim\varphi(v)$.

Now, let $\xi[\mu|\tau]$ be a sentence obtained from a *simple* sentence ξ by replacing τ (in ξ) in exactly one place by μ. Then the remaining axioms of System A are instances of the following schemata:

VI. $\varsigma = \varsigma$.
VII. $\varsigma_1 = \varsigma_2 \rightarrow [\xi \rightarrow \xi[\varsigma_2|\varsigma_1]]$.

Instances of these pertain to the *logic of identity*.

---

17. I use lower-case Greek letters (and sometimes with "*s*" attached), with or without numerical subscripts, as meta-variables: "τ" and "μ" for any terms, singular or plural; "ς" for *singular* terms; "π" for predicates; "φ" and "ψ" for sentences; ξ for *simple* sentences; "*v*" for *singular* variables; and "*vs*" for *plural* variables. The two conditions "τ is substitutable for *v* in φ(*v*)" and "*v* occurs free in φ" used in the specifications of the schemata can be defined in an obvious way for the canonical notation (*v* (or *vs*) occurs *free* in a term τ, if *v* (or *vs*) occurs in τ).

To characterize the theorems of System A on the basis of its axioms, we need an extended notion of universal generalization that is suitable for the canonical notation. Say that a sentence φ is a *plural universal generalization* of a sentence ψ if, and only if, there are (one or more) variables $\tau_1, \tau_2, \ldots, \tau_n$, singular or plural, such that the following holds:

$$\varphi = \ulcorner Q_1\tau_1 Q_2\tau_2 \ldots Q_n\tau_n\psi \urcorner,$$

where $Q_1, Q_2, \ldots, Q_n$ are the singular or plural universal quantifiers that the respective variables are appropriate for. Now, we can give a recursive characterization of the theorems of System A:

> *Theorem of System A*: A sentence φ of the canonical notation is a *theorem of System A* (in symbols, $\vdash_A \varphi$) if, and only if, one of the following conditions holds:
>
> (a) φ is an axiom of *System A*.
> (b) φ is a plural universal generalization of an axiom of *System A*.
> (c) There is a sentence ψ in the canonical notation such that both $\ulcorner \psi \rightarrow \varphi \urcorner$ and ψ are theorems of *System A*.

Then deducibility in System A can be characterized in the usual way:

> *Deducibility in System A*: Let φ be a sentence in the canonical notation and Γ a set of sentences in the notation. Then φ *is deducible from* Γ *in System A* (in symbols, $\Gamma \vdash_A \varphi$) if, and only if, one of the following conditions holds:
>
> (a) φ is a theorem of *System A*.
> (b) φ is a sentence in Γ.
> (c) There is a sentence ψ in the canonical notation such that $\ulcorner \psi \rightarrow \varphi \urcorner$ and ψ are deducible from Γ in *System A*.[18]

18. In presenting the other three systems, I shall specify only their axioms. The theorems of the systems (or deducibility in them) can be characterized just as the theorems of System A (or deducibility in it) is here characterized.

System A is essentially a system that characterizes elementary logic.[19] But it is somewhat stronger than the corresponding system that, couched in elementary notation, characterizes elementary logic. Theorems of System A include some closed sentences that are not logical truths of elementary logic because they are not in elementary notation, such as "$\mathbf{R}(ys, x) \rightarrow \mathbf{R}(ys, x)$", "$\mathbf{R}([j@c], b) \rightarrow \mathbf{R}([j@c], b)$", and "$\forall x \Pi ys[\mathbf{R}(ys, x) \rightarrow \mathbf{R}(ys, x)]$". The first and the second are truth-functional tautologies in the canonical notation; the third a plural universal generalization of the first.[20]

The other three systems are successive extensions of System A. System B has additional axioms dealing with the logic of plural quantifications. System C contains further additions that concern two kinds of predicates: neutral expansions and singular reducts. System D adds axioms that deal with logical properties or relations that rest on two logical predicates: "**S**" (or "to **be-the-same-as**") and "**η**" (or "**is-one-of**").

2.2.1.2 System B

In addition to the axioms of System A, System B includes as axioms instances of four schemata:

VIII. $\varphi(\tau) \rightarrow \Sigma \upsilon s \varphi(\upsilon s)$, where τ is substitutable for υs in $\varphi(\upsilon s)$.
IX. $\Pi \upsilon s[\varphi(\upsilon s) \rightarrow \psi(\upsilon s)] \rightarrow [\Pi \upsilon s \varphi(\upsilon s) \rightarrow \Pi \upsilon s \psi(\upsilon s)]$.
X. $\varphi \rightarrow \Pi \upsilon s \varphi$, where υs does not occur free in φ.
XI. $\Pi \upsilon s \varphi(\upsilon s) \leftrightarrow \sim\Sigma \upsilon s \sim\varphi(\upsilon s)$.[21]

---

19. I have obtained the system by making nominal modifications to the system given in Enderton (1972, 104), which gives a complete characterization of elementary logic.

20. In System A, a plural universal generalization of a theorem is not necessarily a theorem (for example, "$\forall x \mathbf{R}(ys, x) \rightarrow \exists x \mathbf{R}(ys, x)$" is its theorem, but "$\Pi ys[\forall x \mathbf{R}(ys, x) \rightarrow \exists x \mathbf{R}(ys, x)]$" is not), although on account of schemata *III* and *IV*, any universal generalization *in the usual sense* of a theorem of System A is a theorem of the system. The other systems, which have the plural counterparts of the two schemata, overcome this limitation of System A.

21. Recall that "υs" is a meta-variable for plural variables (see note 17).

These schemata, as plural counterparts of schemata *II-V* in System A, characterize the *logic of plural quantifications.*

Thus System B gives proper treatments of, e.g., the following argument:

John and Carol are children who carry Bob.
Therefore, some children carry Bob.

Its premiss and conclusion can be paraphrased into the canonical notation as follows:

(12') $C^N([j@c]) \wedge \mathbf{R}([j@c], b)$.
(13') $\Sigma xs[C^N(xs) \wedge \mathbf{R}(xs, b)]$.

Now the conditional whose antecedent is (12') and whose consequent is (13') is an instance of *VIII*. Thus the argument can be shown to be valid in System B.

Notice that instances of *VIII* include the following:[22]

$[C^N(j) \wedge H^N(j)] \rightarrow \Sigma xs[C^N(xs) \wedge H^N(xs)]$.
$[C^N(j) \wedge \mathbf{R}(j, b)] \rightarrow \Sigma xs[C^N(xs) \wedge \mathbf{R}(xs, b)]$.

Accordingly, paraphrases of the following arguments into canonical notation are also valid on the logic of plurals:

John is a child who is healthy.
Therefore, some children are healthy.

John is a child who carries Bob.
Therefore, some children carry Bob.[23]

---

22. For τ is for any terms, singular or plural.

23. At this stage, this depends on taking the antecedents of the above instances to paraphrase the premisses of the arguments. One might object that they must be paraphrased in the usual way by "$C(j) \wedge H(j)$" and "$C(j) \wedge C_1(j, b)$", respectively. I can meet this objection below, using axioms of System C.

This is the correct result. As I understand plural quantifiers in English, the sentence "Some children are healthy" is true as long as there is at least one child (e.g., John) who is healthy.[24]

On the same understanding, the following arguments (and their paraphrases into the canonical notation) are valid as well:

> *A* child is healthy.
> Therefore, *some* children are healthy.

> *A* child carries Bob.
> Therefore, *some* children carry Bob.

Using instances of *VIII*, we can show them to be valid. Consider the following sentences:

$$[C^{N}(x) \wedge H^{N}(x)] \rightarrow \Sigma xs[C^{N}(xs) \wedge H^{N}(xs)].$$
$$[C^{N}(x) \wedge \mathbf{R}(x, b)] \rightarrow \Sigma xs[C^{N}(xs) \wedge \mathbf{R}(xs, b)].$$

They are instances of *VIII*, and the variable "x" occurs free only in their antecedents. So we can conclude that the following are also its theorems:

$$\exists x[C^{N}(x) \wedge H^{N}(x)] \rightarrow \Sigma xs[C^{N}(xs) \wedge H^{N}(xs)].$$
$$\exists x[C^{N}(x) \wedge \mathbf{R}(x, b)] \rightarrow \Sigma xs[C^{N}(xs) \wedge \mathbf{R}(xs, b)].$$ [25]

In drawing this conclusion, I have appealed to the following principle:

> Suppose that a conditional $\ulcorner \varphi \rightarrow \psi \urcorner$ is deducible from Γ in a system containing A and that a singular variable v occurs free neither in ψ nor in sentences in Γ. Then $\ulcorner \exists v \varphi \rightarrow \psi \urcorner$ is also deducible from Γ in the system.

24. Thus it can be elaborated as "*One or* more children are healthy." Similarly, "Some children carry Bob" can be elaborated as "One or more children carry Bob."

25. In general, instances of the schema "$\exists v\varphi(v) \rightarrow \Sigma vs\varphi(vs)$" are theorems of System B.

This principle holds because System A includes instances of schemata *III-V* as well as truth-functional tautologies. The proof of this shows the critical role of the schemata.[26] Similarly, the role of schemata *IX-XI* in System B can be taken to lie in yielding the plural counterpart of the above principle:

> Suppose that a conditional $\ulcorner\varphi\rightarrow\psi\urcorner$ is deducible from Γ in a system containing B and that a plural variable υs occurs free neither in ψ nor in sentences in Γ. Then $\ulcorner\Sigma\upsilon s\varphi\rightarrow\psi\urcorner$ is also deducible from Γ in the system.

2.2.1.3 System C

In addition to axioms of System B, System C contains axioms pertaining to the neutral expansions and singular reducts. Let $\xi^S(\tau)$ and $\xi^N(\tau)$ be sentences whose predicates are the *singular reduct* and the *neutral expansion*, respectively, of the predicate of the simple sentence $\xi(\tau)$ with respect to the argument place filled in with τ. Then the following schema gives a complete characterization of singular reducts:[27]

XII. $\xi^S(\varsigma)\leftrightarrow\xi(\varsigma)$.[28]

And the following three schemata pertain to neutral expansions:

XIII. $\xi^N(\varsigma)\leftrightarrow\xi(\varsigma)$.
XIV. $\xi^N([\tau@\mu])\leftrightarrow[\xi^N(\tau)\wedge\xi^N(\mu)]$.
XV. $\Sigma\upsilon s\xi^N(\upsilon s)\rightarrow\exists v\xi^N(v)$.

Instances of XII-XV are axioms of System C.

---

26. In Gentzen-style formulations of elementary logic, the principle is expressed as the rule of elimination of the singular existential quantifier.

27. In giving axioms dealing with singular reducts or neutral expansions, I take into account only the reduct or expansion obtained by varying one argument place of a base predicate. The reduct or expansion with respect to more than one argument place can be considered as resulting from iteration of varying one argument place.

28. Recall that ς is for singular terms only; but $\ulcorner\xi^S(\tau)\urcorner$ is not well-formed if τ is a plural term.

Notice that it is critical to formulate schemata *XIV* and *XV* in this restricted way with reference to neutral expansions. For "$\mathbf{R}([j@c], b) \leftrightarrow [\mathbf{R}(j, b) \wedge \mathbf{R}(c, b)]$" cannot be included as a theorem of a logical system. The sentences flanking the biconditional are paraphrases of "John and Carol carry Bob" and "John carries Bob and Carol carries Bob", respectively, but these sentences are not logically equivalent. Similarly, "$\Sigma xs\mathbf{R}(xs, b) \rightarrow \exists x\mathbf{R}(x, b)$" cannot be a theorem of a logical system; its antecedent and consequent paraphrase "There are some things (e.g., John and Carol) that carry Bob" and "There is something (e.g., John) that carries Bob", respectively.

Schema *XIV* codifies the distributivity of neutral expansions over the term connective "@" (or "**and**"). The schema, aided with *XIII*, accounts for the logical equivalence between, for example, "John is a child *and* Carol is a child" (sentence (5)) and "John *and* Carol are children" (sentence (6)). Consider their paraphrases into the canonical notation. The natural paraphrase of (6) is "$C^N([j@c])$"; and the usual paraphrase of (5) is "$[C(j) \wedge C(c)]$". Now, the following are instances of schemata *XIV* and *XIII*: "$C^N([j@c]) \leftrightarrow [C^N(j) \wedge C^N(c)]$", "$[C^N(j) \leftrightarrow C(j)]$", and "$[C^N(c) \leftrightarrow C(c)]$". So "$C^N([j@c]) \leftrightarrow [C(j) \wedge C(c)]$" is a theorem of System C.

Notice that this explanation of the logical equivalence relies on the usual paraphrase "$[C(j) \wedge C(c)]$" of sentence (5). But the above paraphrase of sentence (6) might yield a consideration in favor of another paraphrase of sentence (5): "$[C^N(j) \wedge C^N(c)]$"; because (5) and (6), it seems, contain the same predicate, if of different forms, their straightforward paraphrases must also contain the same predicate. Given this paraphrase of (5), Schema *XIII*, of which "$C^N([j@c]) \leftrightarrow [C^N(j) \wedge C^N(c)]$" is an instance, suffices to account for the logical equivalence between (5) and (6).

So we can see that the role of the two instances of *XIII* in the first explanation is to yield the logical equivalence between the two candidates for paraphrasing the same English sentence (5). I think this is the correct result–whatever is the right answer to the further question of which candidate, if any, is *the* correct paraphrase of (5).[29]

---

29. My present view is: the second candidate, "$[C^N(j) \wedge C^N(c)]$", is the more *natural* paraphrase, on account of the above consideration; but both are correct paraphrases. (The same sentence can be paraphrased by more than one sentence in the canonical notation, as the "neither . . . nor" constructions in English can be paraphrased in two different ways into elementary notation.)

Similarly, we can see that the role of *XII* lies in relating two plausible candidates for paraphrasing, e.g., sentence (9), "John carries Bob." Its usual paraphrase is "$C_1(j, b)$", where "C_1" (or "carries") is a singular predicate available in elementary notation. Another possible paraphrase, which conforms to the view that the sentence has the same predicate that occurs in, e.g., "John and Carol carry Bob", is "$\mathbf{R}(j, b)$." Now, recall that the predicates "C_1" and "$\mathbf{R}$" are introduced into the canonical notation on the basis of the same English predicate that occurs in sentence (9): "$\mathbf{R}$" is introduced as a rendering of the (putative) predicate "to carry" that can take singular and plural forms; and "C_1" is introduced as a rendering of the singular form (or "carries") of the predicate. For this reason, I have identified the singular predicate "C_1" as the singular reduct of "$\mathbf{R}$" (and proposed to write it as "$\mathbf{R}^{S}$").[30] Thus, the biconditional "$C_1(j, b)\leftrightarrow\mathbf{R}(j, b)$" is an instance of *XII*.

Schema *XV*, too, helps to clarify the logical equivalence between two candidates for paraphrasing, for example, "There are some children." Its straightforward paraphrase is "$\Sigma xs C^{N}(xs)$"; its usual paraphrase is "$\exists x C(x)$", which can be shown to be logically equivalent with "$\exists x C^{N}(x)$" using an instance of *XIII*. Now, "$\Sigma xs C^{N}(xs)\rightarrow\exists x C^{N}(x)$" is an instance of *XV* and its converse is a theorem of System B.

Notice, however, that schema *XV*, even aided with the other schemata available thus far, fails to explain the logical equivalence between two candidates for paraphrasing, e.g., sentence (13), "Some children are healthy": the straightforward paraphrase "$\Sigma xs[C^{N}(xs)\wedge \boldsymbol{H}^{N}(xs)]$" (sentence (13')) and the usual paraphrase "$\exists x[C(x)\wedge H(x)]$" (sentence (2')). We cannot resolve this problem by directly generalizing *XV*. Consider, e.g., the following:

$$\Sigma xs[\sim C^{N}(xs)\wedge\sim H^{N}(xs)].$$
$$\exists x[\sim C^{N}(x)\wedge\sim H^{N}(x)].$$

The first of these sentences does not imply the second. To see this, suppose that there is only one thing that is not a child (viz., Bill), that there is only one thing that is not healthy (viz., John), and that the former is not the latter. In that case, the first sentence would be true on account of Bill and John (taken together), but

30. The same observation applies to the singular predicate "C" (or "is-a-child") and its neutral expansion "C^{N}" (or "to be-child[ren]"). The singular reduct of "C^{N}" can be identified with "C".

the second would be false.[31] We can overcome the limitation of *XV* with the following:

XVI. $\xi^{N}(\tau) \leftrightarrow \forall \nu[\eta(\nu, \tau) \rightarrow \xi(\nu)]$, where ν does not occur free in $\xi(\tau)$.

This schema yields the desired result on the paraphrases of, e.g., sentence (13). Moreover, we can use it to derive all the above schemata, *XIII-XV*, that pertain to neutral expansions; *XVI* can be shown to yield a complete characterization of neutral expansions in terms of their base predicate together with the predicate "**η**" (or "**is-one-of**"). But this pleasing result rests on the logic of this special predicate, which lies beyond the scope of System C. Let me now turn to the next system, which can deal with its logic.

2.2.1.4 System D

In addition to the predicate "=" (or "is-identical-with") available in elementary notation, recall, the canonical notation marks two other predicates, "**S**" (or "to **be-the-same-as**") and "**η**" (or "**is-one-of**"), as logical predicates. System D contains axioms that pertain to these two predicates. I shall formulate the additional axioms, while giving English sentences that can be paraphrased by the axioms, and present straightforward consequences of the axioms.

Axioms: The additional axioms are instances of eight schemata. The schemata can be seen to belong to four groups. First, schema *XVI*, mentioned above, characterizes neutral expansions:

XVI. $\xi^{N}(\tau) \leftrightarrow \forall \nu[\eta(\nu, \tau) \rightarrow \xi(\nu)]$, where ν does not occur free in $\xi(\tau)$.

An English sentence that can be paraphrased by an instance of this schema is: "John and Carol are children if, and only if, every one of them is a child."

Second, two schemata pertain to the logic of the predicate "**S**":

---

31. This does not yield a counterexample to the logical equivalence between "Some non-children are not healthy" and "A non-child is not healthy." For "to be non-children" cannot be identified with "not to be children" (i.e., the negation of the neutral expansion of "is a child"). I think the predicate "to be non-children" is the 'neutral expansion' of the complex singular predicate "is not a child", the complement of "is a child".

XVII. $\mathbf{S}(\tau, \mu) \rightarrow [\xi \rightarrow \xi[\mu|\tau]]$.[32]

XVIII. $\forall \nu[\boldsymbol{\eta}(\nu, \tau) \leftrightarrow \boldsymbol{\eta}(\nu, \mu)] \rightarrow \mathbf{S}(\tau, \mu)$, where ν does not occur free in $\mathbf{S}(\tau, \mu)$.

The paraphrase of the following sentence is an instance of *XVII*: "If they *are* (or are the same as) John and Carol, then they carry Bob just in case John and Carol carry it." Notice that the following sentence amounts to a plural universal generalization of the instance: "If there are some things that are John and Carol, then they carry Bob just in case John and Carol carry Bob." Instances of *XVIII* include the natural paraphrase of "If something is one of them just in case it is one of John and Carol, then they *are* (or are the same as) John and Carol."

Third, four schemata concern the logical behavior of plural terms with respect to predicate "$\boldsymbol{\eta}$":

XIX. $\boldsymbol{\eta}(\varsigma, [\tau @ \mu]) \leftrightarrow [\boldsymbol{\eta}(\varsigma, \tau) \vee \boldsymbol{\eta}(\varsigma, \mu)]$.

XX. $\boldsymbol{\eta}(\varsigma_1, [\varsigma_2 @ \tau]) \leftrightarrow [\varsigma_1 = \varsigma_2 \vee \boldsymbol{\eta}(\varsigma_1, \tau)]$.[33]

XXI. $\boldsymbol{\eta}(\varsigma_1, \varsigma_2) \rightarrow \varsigma_1 = \varsigma_2$.

XXII. $\exists \nu \boldsymbol{\eta}(\nu, \upsilon s)$.

Sentences paraphrased by their instances are: "Tully is one of these and those if, and only if, he is one of these or he is one of those" (*XIX*); "Tully is one of Cicero and those if, and only if, either he is Cicero or he is one of those" (*XX*); the quasi-English sentence "If Cicero is one of Tully, then Cicero is Tully" (*XXI*); "There is something that is one of them" (*XXII*), which is contained in, e.g., "If there are some things, then there is something that is one of them."

The last schema pertains to the condition for there to be some (one or more) things:

XXIII. $\exists \nu \varphi(\nu) \rightarrow \Sigma \upsilon s \forall \nu[\boldsymbol{\eta}(\nu, \upsilon s) \leftrightarrow \varphi(\nu)]$, where υs does not occur free in $\varphi(\nu)$.

Its instances include straightforward paraphrases of the following:

---

32. Notice that if τ is a singular term that occurs in a singular argument place of the predicate of ξ while μ is a plural term, the schema does not have a well-formed instance. See p. 39 for the definition of $\xi[\mu|\tau]$.

33. Notice that τ and μ in *XIX* and *XX* are *not* required to be plural terms.

If there is a child in the Common Room, then there are some things such that something is one of them if and only if it is a child in the Common Room.
If a child carries Bob, then there are some things such that something is one of them if and only if it is a child and carries Bob.
If something is loved by a child, then there are some things such that something is one of them if and only if it is loved by a child.

Key Results: To draw some key results of the axioms, it is useful to begin with schemata *XIX-XXII*. Notice that "τ" in *XIX* can be replaced by a singular term. This leads to a mild surprise. Schemata *XIX* and *XX* yield "$\boldsymbol{\eta}(\varsigma, \varsigma)$". This together with *XXI* yields "$\boldsymbol{\eta}(\varsigma_1, \varsigma_2) \leftrightarrow \varsigma_1 = \varsigma_2$". So instances of the two schemata are theorems of System D:

Theorem A. (a) $\boldsymbol{\eta}(\varsigma, \varsigma)$; and (b) $\boldsymbol{\eta}(\varsigma_1, \varsigma_2) \leftrightarrow \varsigma_1 = \varsigma_2$.

This result helps to reduce the logic of neutral expansions to *XVI*: $\xi^N(\tau) \leftrightarrow \forall \nu[\boldsymbol{\eta}(\nu, \tau) \rightarrow \xi(\nu)]$. The result together with *XVI* yields *XIII*: $\xi^N(\varsigma) \leftrightarrow \xi(\varsigma)$. The next two schemata, *XIV* and *XV*, are also reducible to *XVI*. Given *XVI*, *XIX* yields *XIV*: $\xi^N([\tau @ \mu]) \leftrightarrow [\xi^N(\tau) \wedge \xi^N(\mu)]$; and *XXII* yields *XV*: $\Sigma \upsilon s \xi^N(\upsilon s) \rightarrow \exists \nu \xi^N(\nu)$.

We can see that schema *XVI* goes beyond the three schemata reduced to it. For example, the schema, combined with *XXII*, yields the following:

$$\Sigma \upsilon s[\xi_1{}^N(\upsilon s) \wedge \xi_2{}^N(\upsilon s)] \rightarrow \exists \nu[\xi_1{}^N(\nu) \wedge \xi_2{}^N(\nu)].$$

Now, an instance of this schema can help to account for the logical equivalence between two possible paraphrases of "Some children are healthy": "$\Sigma xs[C^N(xs) \wedge H^N(xs)]$" and "$\exists x[C(x) \wedge H(x)]$." At the same time, *XVI* helps to clarify the counterexample discussed above to the logical equivalence between "$\Sigma xs[\sim C^N(xs) \wedge \sim H^N(xs)]$" and "$\exists x[\sim C(x) \wedge \sim H(x)]$."

Let us now consider the schemata *XVII* and *XVIII* that pertain to the predicate "**S**". They lead to two main results. First, they lead to the natural characterization of "**S**":

Theorem B. $\forall \nu[\boldsymbol{\eta}(\nu, \tau) \leftrightarrow \boldsymbol{\eta}(\nu, \mu)] \leftrightarrow \mathbf{S}(\tau, \mu)$, where ν does not occur free in $\mathbf{S}(\tau, \mu)$.

One direction of the biconditional is *XVIII*. The other direction results from an instance of *XVII*: $\mathbf{S}(\tau, \mu) \rightarrow [\eta(\nu, \tau) \leftrightarrow \eta(\nu, \mu)]$.[34] Second, a full parallel can be drawn between the neutral predicate "**S**" and the predicate "=" for identity. Recall the two schemata that characterize the logic of identity:

VI. $\varsigma = \varsigma$.
VII. $\varsigma_1 = \varsigma_2 \rightarrow [\xi \rightarrow \xi[\varsigma_2|\varsigma_1]]$.

Schema *XVII* is precisely the neutral counterpart of *VII*. And an 'instance' of *XVIII*, viz. the schema "$\forall\nu[\eta(\nu, \tau) \leftrightarrow \eta(\nu, \tau)] \rightarrow \mathbf{S}(\tau, \tau)$", yields the neutral counterpart of (6):

Theorem C. $\mathbf{S}(\tau, \tau)$.

Thus the logic of the predicate "**S**" draws an exact parallel to that of the predicate "=".

What remains for discussion is *XXIII*: $\exists\nu\varphi(\nu) \rightarrow \Sigma\upsilon s\forall\nu[\eta(\nu, \upsilon s) \leftrightarrow \varphi(\nu)]$. This schema relates to, for example, the following arguments:

There are some children.
Therefore, there are some children of whom every child is one.

Some children are healthy.
Therefore, there are some healthy children of whom every healthy child is one.[35]

---

34. We can bind the free variable ν in the consequent of this schema with the singular universal quantifier and the result is still a theorem schema on account of schemata *III* and *IV*.

35. The conclusions of these arguments can be shown to be logically equivalent with the following, respectively: "There are some things such that something is one of them if and only if it is a child" and "There are some things such that something is one of them if and only if it is a child and is healthy." The logical equivalence rests on *XVI*.

I think it should be clear that these are valid arguments. In order to account for their validity by considering their paraphrases into the canonical notation, we need to appeal to instances of *XXIII*, such as the following:

$$\exists x C^{N}(x) \rightarrow \Sigma ys \forall x[\eta(x, ys) \leftrightarrow C^{N}(x)].$$

So instances of the schema are included as axioms of System D.[36] The schema, correctly understood, must strike one as the natural one to adopt as a schema of logical truths. To distinguish it clearly from its many variations, most of which are clearly unacceptable, however, it is necessary to pay attention to its fine details.

First, the antecedent "$\exists v\varphi(v)$" of the schema is essential. Taking the consequent of the schema as a schema of logical truths would yield the wrong result that "There are some children", for example, is a logical truth on the logic of plurals.

Second, the antecedent "$\exists v\varphi(v)$" cannot be replaced with "$\Sigma vs\varphi(vs)$". Doing so would lead to the result that the following arguments, for example, are valid:

> *Some* child*ren* carry a piano.
> Therefore, there are some children, every one of them carries a piano, and every child who carries a piano is one of them.

But this is wrong. Consider the familiar case in which John and Carol, two children, collaborate and carry a piano while no child alone carries a piano. In that case, the premiss would be true while the conclusion being false.

Third, notice that no restriction whatsoever is placed on sentences that can replace "$\varphi(v)$" in the schema. Replacing it with, e.g., "$C^{N}(x) \wedge \mathbf{R}(x, b)$" yields a

---

36. One consequence of the schema is worthy of special notice on account of its connection to the famous theorem of Cantor's:

$$(\Pi vs \exists v\varphi(v, vs) \wedge \forall v \Pi vs_1 \Pi vs_2[\varphi(v, vs_1) \wedge \varphi(v, vs_2) \rightarrow \mathbf{S}(vs_1, vs_2)]) \rightarrow \forall v_1 \forall v_2 v_1 = v_2.$$

This schema, whose instances are logically true on the logic of plurals, can be shown to underlie the theorem–although the theorem itself is not logically true on the logic.

legitimate instance of the schema. So the schema can account for the validity of, e.g., the following argument:

> A child carries a piano.
> Therefore, there are some children, every one of them carries a piano, and every child who carries a piano is one of them.

The validity of this argument is immune to the above case given as a counterexample to the validity of the previous argument. In the case in question, the premiss of the present argument, as well as its conclusion, would be false.[37]

Finally and perhaps most importantly, it is essential to use the combination of the plural existential quantifier and the singular universal quantifier in the consequent of the schema. Replacing the plural quantifier and the associated variable with their singular counterparts yields a schema that is unacceptable:

$$\exists v\varphi(v) \rightarrow \exists v_1 \forall v[\eta(v, v_1) \leftrightarrow \varphi(v)].$$

Because $\eta(x, y)$ if and only if $x=y$, the instance of this schema obtained by replacing "$\varphi(v)$" with "$\exists z{\sim}x=z$" (and "v" with "x") is logically equivalent with "${\sim}\exists x \exists z {\sim} x=z$" (on elementary logic). Clearly, however, this sentence is not a logical truth.[38]

2.2.2 *The Logic of Plurals: Model Theory*[39]

In this subsection, I give a complete characterization of the logic of plurals in model-theoretic terms. That is to say, I define the two notions:

---

37. Consider another case: Chelsea is a child who (alone) carries a piano; Tom is also a child who does so; two other children, John and Carol, carry Bob (together) but none of them does so alone; and there is no other child who carries a piano (alone or together with others). In this case, the conclusion, too, would be true; *Chelsea and Tom* are children, each of *them* carries a piano, and every child who carries a piano is one of *them*.

38. Parmenides of Elea would disagree.

39. This subsection is especially sketchy and cryptic. I hope to write an expanded version shortly.

> φ *is a model-theoretic truth in* $\mathscr{L}$ (in symbols, $\models^{\mathscr{L}} \varphi$)
> φ *is a model-theoretic consequence in* $\mathscr{L}$ *of* Γ (in symbols, $\Gamma \models^{\mathscr{L}} \varphi$)[40]

where $\mathscr{L}$ is a language in the canonical notation, φ a *closed* sentence in $\mathscr{L}$, and Γ a set of *closed* sentences in $\mathscr{L}$, so that the following hold:

> A closed sentence φ is a model-theoretic truth (in $\mathscr{L}$) if, and only if, φ is *logically true* (in $\mathscr{L}$).
>
> φ is a model-theoretic consequence (in $\mathscr{L}$) of Γ if, and only if, φ is a *logical consequence* (in $\mathscr{L}$) of the sentences in Γ.[41]

To define the desired notions it is necessary to introduce the notion of *structure* (or *model*) suitable for languages in the canonical notation. A *structure* M *for* a language $\mathscr{L}$ in the canonical notation is a pair $<D, f>$ that satisfies the following conditions, where $D^+=\mathcal{P}(D)\backslash\{\varnothing\}$:[42]

(A1) D is a non-empty set.

(A2) f is a function that satisfies the following conditions:

- (a) $f(\varsigma)\varepsilon D$, if ς is a singular constant.
- (b) $f(=)=\{<x, x>: x\varepsilon D\}$.
- (c) $f(\mathbf{S})=\{<x, x>: x\varepsilon D^+\}$.
- (d) $f(\eta)=\{<x, y>: x\varepsilon D\wedge y\varepsilon D^+\wedge x\varepsilon y\}$.
- (e) $f(\pi) \subseteq D_1 x D_2 x \ldots x D_n$, if π is an n-place predicate (where D_i is D if the i-th argument place of π is singular; and D^+ otherwise).
- (f) $<x_1, x_2, \ldots, x_n>\varepsilon f(\pi^{S(i)})$ iff $<x_1, \ldots, \{x_i\}, \ldots, x_n>\varepsilon f(\pi)$, if π is n-place.
- (g) $<x_1, x_2, \ldots, x_n>\varepsilon f(\pi^{N(i)})$ iff $\forall y_i\varepsilon x_i <x_1, \ldots, y_i, \ldots, x_n>\varepsilon f(\pi)$, if π is n-place.

---

40. I leave out the superscript $\mathscr{L}$ and write "$\models\varphi$" and "$\Gamma\models\varphi$", respectively, if the language in question is clear from the context.

41. Γ can be the empty set. In that case, I say that φ is a logical consequence of the sentences in Γ if, and only if, φ is a logical truth.

42. $\mathcal{P}(D)$ is the power set of D; $\varnothing$ the empty set.

And say that a function s whose domain is the set of variables of language $\mathscr{L}$ is an *assignment appropriate for* a structure M (for $\mathscr{L}$), if the following hold:

(B1) $s(v)\varepsilon D$.
(B2) $s(vs)\varepsilon D^{+}$.

Now, let $\mathscr{L}$ be a language, M a structure for $\mathscr{L}$, and s an assignment appropriate for M. Then *the value of* a term τ (in $\mathscr{L}$) *in* M *with respect to* s (in symbols, $\|\tau\|^{M}_{s}$) can be defined as follows:

(C1) $\|\tau\|^{M}_{s}=f(\tau)$, if τ is a singular constant.
(C2) $\|\tau\|^{M}_{s}=s(\tau)$, if τ is a variable.
(C3) $\|\upsilon\|^{M}_{s}=(\|\tau_1\|^{M}_{s})' \cup (\|\tau_2\|^{M}_{s})'$, if $\upsilon=\ulcorner[\tau_1@\tau_2]\urcorner$.

In (C3), $(\|\tau_i\|^{M}_{s})'$ is $\{\|\tau_i\|^{M}_{s}\}$ if τ_i is a singular term, and $\|\tau_i\|^{M}_{s}$ if τ_i is a plural term (for i=1 or 2) (in general, $(\|\tau\|^{M}_{s})'$ s $\{\|\tau\|^{M}_{s}\}$ if τ is a singular term, and $\|\tau\|^{M}_{s}$ if τ is a plural term). Next, we can define the satisfaction relation

φ *is satisfied by* s *in* M (in symbols, $M\models\varphi[s]$)

where φ is a sentence (in $\mathscr{L}$), open or closed. If x is in D or D^{+}, τ and υ variables (in $\mathscr{L}$), and s an assignment, then let $s(\tau|x)$ be an assignment such that $s(\tau|x)(\upsilon)$ is x if $\upsilon=\tau$, and $s(\upsilon)$ otherwise. Then the satisfaction relation can be defined recursively by the following conditions:

(D1) $M\models\varsigma_1=\varsigma_2[s]$ iff $\|\varsigma_1\|^{M}_{s}=\|\varsigma_2\|^{M}_{s}$.
(D2) $M\models S(\tau, \upsilon)[s]$ iff $(\|\tau\|^{M}_{s})'=(\|\upsilon\|^{M}_{s})'$.
(D3) $M\models\eta(\varsigma, \tau)[s]$ iff $\|\varsigma\|^{M}_{s}\,\varepsilon\,(\|\tau\|^{M}_{s})'$.
(D4) $M\models\pi(\tau_1, \tau_2, \ldots, \tau_n)[s]$ iff $<(\|\tau_1\|^{M}_{s})', (\|\tau_2\|^{M}_{s})', \ldots, (\|\tau_n\|^{M}_{s})'> \varepsilon\, f(\pi)$, if π is an n-place predicate (where $(\|\tau\|^{M}_{s})'$ is $\{\|\tau\|^{M}_{s}\}$ if τ is a singular term while the i-th argument place of π is neutral; and $\|\tau\|^{M}_{s}$ otherwise).
(D5) $M\models(\varphi\wedge\psi)[s]$ iff $M\models\varphi[s]$ and $M\models\psi[s]$.
(D6) $M\models\sim\varphi[s]$ iff it is not the case that $M\models\varphi[s]$.
(D7) $M\models\forall v\varphi(v)[s]$ iff, for anything x in D, $M\models\varphi[s(v|x)]$.
(D8) $M\models\Pi vs\varphi(vs)[s]$ iff, for anything x in D^{+}, $M\models\varphi[s(vs|x)]$.

Finally, we can complete the definition of the model-theoretic truth and consequence in the usual way. Let φ be a closed sentence in $\mathscr{L}$, and Γ a set of closed sentences in $\mathscr{L}$. Then

(E1) φ *is satisfied in M* (in symbols, $M \models \varphi$) iff, for any assignment *s* appropriate for *M*, $M \models \varphi[s]$.

(E2) *M satisfies the sentences in* Γ (in symbols, $M \models \Gamma$) iff, for any sentence φ in Γ, $M \models \varphi$.

(E3) $\Gamma \models \varphi$ iff, for any structure *M* for $\mathscr{L}$, if $M \models \Gamma$ then $M \models \varphi$.

(E4) $\models \varphi$ iff $\varnothing \models \varphi$.

Given the model-theoretic characterization, it can be shown that systems A-D are approximations to the logic of plurals:

> *Soundness*: Let φ be a closed sentence (in $\mathscr{L}$), Γ a set of closed sentences (in $\mathscr{L}$), and S one of the systems A-D. Then if $\Gamma \vdash_S \varphi$, then $\Gamma \models \varphi$.[43]

It can also be shown that the logic of plurals is a *conservative extension* of elementary logic. Given a language $\mathscr{L}$ in the canonical notation, it is straightforward to define the *elementary fragment of* $\mathscr{L}$ (in symbols, $E(\mathscr{L})$) that is a language in elementary notation. Then if φ is a closed sentence in $E(\mathscr{L})$ and Γ a set of closed sentences in $E(\mathscr{L})$, the following hold:

> $\Gamma \models^{\mathscr{L}} \varphi$ if, and only if, φ is a logical consequence (in $E(\mathscr{L})$) of the sentences in Γ *on elementary logic*.
>
> $\models^{\mathscr{L}} \varphi$ if, and only if, φ is logically true (in $E(\mathscr{L})$) *on elementary logic*.

Thus the logic of plurals verifies that the following, for example, are logical truths:

(14) If John and Carol are children who carry a piano, then some children carry a piano.

(15) If Ezra and Thomas are critics who admire only each other, then some critics admire only one another.

---

43. But the converse does not hold.

Their paraphrases into the canonical notation are instances of schema *VIII*: $\varphi(\tau) \rightarrow \Sigma \upsilon s \varphi(\upsilon s)$. By contrast, the following, for example, are *not* logical truths on the logic of plurals:

(16) If Ezra and Thomas are critics who admire only each other, then there is something (e.g., a set) every member of which is a critic and admires only its other member.

(17) If some critics admire only one another, then there is something (e.g., a set) every member of which is a critic and admires only its other member.

Sentence (16) can be paraphrased by a sentence in elementary notation that is not a logical truth on elementary logic.[44] So the paraphrase is not a logical truth (in the canonical notation) on the logic of plurals, either. Then, because (15) is a logical truth while (16) is not on the logic of plurals, (17) is not a logical truth on the logic. I think this is the right result on, e.g., (16) and (17).[45] They are no more logically true than, e.g., the following:

(18) If John and Carol are children, then there is something (e.g., a set) of which John, as well as Carol, is a member.

(19) If some children are healthy, then there is something (e.g., a set) every member of which is a child who is healthy.

Thus the logic of plurals can maintain the distinction between (14)-(15), on the one hand, and (16)-(19), on the other, in the natural way. I think this is one of its major virtues.

---

44. For the predicate "is a member of" (or "ε") for membership is a singular predicate available in elementary notation.

45. See Chapter 1 for an argument for this.

Appendix: Partial Axiomatization of the Logic of Plurals

The plural universal generalizations of the following schemata are logical truths in the languages of plurals in the canonical notation (and the logical truths are closed under *Modus Ponens*):[46]

I.	Truth-functional tautologies.
II.	$\varphi(\varsigma)\rightarrow\exists\nu\varphi(\nu)$, where ς is substitutable for ν in $\varphi(\nu)$.
III.	$\forall\nu[\varphi(\nu)\rightarrow\psi(\nu)] \rightarrow [\forall\nu\varphi(\nu)\rightarrow\forall\nu\psi(\nu)]$.
IV.	$\varphi\rightarrow\forall\nu\varphi$, where ν does not occur free in φ.
V.	$\forall\nu\varphi(\nu) \leftrightarrow {\sim}\exists\nu{\sim}\varphi(\nu)$.
*VI.**	$\varsigma=\varsigma$.
VII.	$\varsigma_1=\varsigma_2\rightarrow(\xi\rightarrow\xi[\varsigma_2\vert\varsigma_1])$.
VIII.	$\varphi(\tau)\rightarrow\Sigma\upsilon s\varphi(\upsilon s)$, where τ is substitutable for υs in $\varphi(\upsilon s)$.
IX.	$\Pi\upsilon s[\varphi(\upsilon s)\rightarrow\psi(\upsilon s)] \rightarrow [\Pi\upsilon s\varphi(\upsilon s)\rightarrow\Pi\upsilon s\psi(\upsilon s)]$.
X.	$\varphi\rightarrow\Pi\upsilon s\varphi$, where υs does not occur free in φ.
XI.	$\Pi\upsilon s\varphi(\upsilon s) \leftrightarrow {\sim}\Sigma\upsilon s{\sim}\varphi(\upsilon s)$.
XII.	$\xi^{S}(\varsigma)\leftrightarrow\xi(\varsigma)$.
*XIII.**	$\xi^{N}(\varsigma)\leftrightarrow\xi(\varsigma)$.
*XIV.**	$\xi^{N}([\tau@\mu])\leftrightarrow[\xi^{N}(\tau)\wedge\xi^{N}(\mu)]$.
*XV.**	$\Sigma\upsilon s\xi^{N}(\upsilon s)\rightarrow\exists\nu\xi^{N}(\nu)$.
XVI.	$\xi^{N}(\tau)\leftrightarrow\forall\nu[\boldsymbol{\eta}(\nu, \tau)\rightarrow\xi(\nu)]$, where ν does not occur free in $\xi(\tau)$.
XVII.	$\mathbf{S}(\tau, \mu)\rightarrow(\xi\rightarrow\xi[\mu\vert\tau])$.
XVIII.	$\forall\nu[\boldsymbol{\eta}(\nu, \tau)\leftrightarrow\boldsymbol{\eta}(\nu, \mu)] \rightarrow \mathbf{S}(\tau, \mu)$, where ν does not occur free in $\mathbf{S}(\tau, \mu)$.
XIX.	$\boldsymbol{\eta}(\varsigma, [\tau@\mu]) \leftrightarrow [\boldsymbol{\eta}(\varsigma, \tau)\vee\boldsymbol{\eta}(\varsigma, \mu)]$.

46. See note 17 for the use of meta-variables. The schemata with the superscripted "*" can be derived from the others. (*VI* can be derived from *XVIII*).

*XX.** $\boldsymbol{\eta}(\varsigma_1, [\varsigma_2@\tau]) \leftrightarrow [\varsigma_1=\varsigma_2 \vee \boldsymbol{\eta}(\varsigma_1, \tau)]$.

XXI. $\boldsymbol{\eta}(\varsigma_1, \varsigma_2) \rightarrow \varsigma_1=\varsigma_2$.

XXII. $\exists\nu\boldsymbol{\eta}(\nu, \upsilon s)$.

XXIII. $\exists\nu\varphi(\nu) \rightarrow \Sigma\upsilon s\forall\nu[\boldsymbol{\eta}(\nu, \upsilon s)\leftrightarrow\varphi(\nu)]$, where υs does not occur free in $\varphi(\nu)$.

3. Is Two a Property?

John and Carol sit next to each other and sing together. They are Jackie's only children. John is not Carol; John is a boy, Carol a girl. So they are *two* children. What is it for them to be *two* children? It is for them to instantiate a *property*, namely, *being two children*. My aim in this chapter is to clarify, defend, and develop this answer.

3.1 Why Property Two?

In this section, I present some grounds for the view that being two children, for example, is a property. To do so, I clarify the content of this view by relating it to the issue of giving proper analysis of some facts, such as the fact that John and Carol are two children (in 3.1.1), and I explain some puzzling features of the facts (in 3.1.2) and argue against the standard analysis of the facts, which can be seen to be based on a strategy of explaining the features away (in 3.1.3). This prepares for the presentation in the following sections of my anti-reductionist analysis, which relates those puzzling features to characteristics of the properties that constitute those facts.

3.1.1 What Is It to Be a Property?

Consider the standard analysis of, for example, the following fact:

1. John is a child.

On the analysis, fact *I* consists in *John*'s instantiation of *being a child.* Notice how this analysis is tied to the view that being a child is a property. To accept the existence of this property is to entertain it as the *predicative part* of such facts as *I* by applying the standard analysis to them (and *vice versa*). Similarly, the view that *being two children* is a property is closely related to an analysis of such facts as the following:

II. John and Carol are two children.

To take this view is to extend the standard analysis of fact *I* to, e.g., fact *II*. Doing so results in the view that this fact, too, can be decomposed into two primary parts:

(a) The *subject part*: John and Carol;
(b) The *predicative part*: being two children.

The fact, on this view, consists in the instantiation of being two children by John and Carol. It is this analysis of, e.g., fact *II* that I propose and defend in this chapter.

Notice that the analysis of fact *II* in terms of the property of being two children conforms to the *prima facie* structural parallel between facts *I* and *II*. Thus it provides the metaphysical framework for a semantic account faithful to the apparent syntactic structures of the English sentences used to state the facts:

(1) John is a child.
(2) John and Carol are two children.

I think this is a pleasing feature of the analysis. But I consider it only a welcome by-product. My primary motivation for accepting the analysis is not the desire to prepare for a semantic account parallel to the natural syntactic account of sentences of, e.g., English. Let me explain.

Compare sentences (1) and (2) with the following:

(3) Either John or Carol is a boy.

I am inclined to think that it is the natural, and correct, view about English as it stands that this sentence has the same overall syntactic structure as sentences (1) and (2).[1] But I am not inclined to embrace an account of the fact stated by sentence (3), i.e., the following fact, simply because it helps to draw a parallel between the structures of that fact and fact *I* (or *II*):

III. Either John or Carol is a boy.

Rather, I shall reject any account that does so by postulating an object above and beyond John and Carol (e.g., a 'disjunctive object' that corresponds to the term "either John or Carol") as the subject part of fact *III*. Fact *III* does not consist in the instantiation of the property of *being a boy* by such an object, which is neither John nor Carol. It is a fact that concerns John and Carol, but nothing (i.e., no *thing*) else. I think this helps to clarify the idea that guides my analysis of fact *II*: roughly speaking, just as fact *I* is a fact about John (and about John *only*), so is fact *II* a fact about John and Carol (and about John and Carol *only*). On this view, facts *I-III* are all facts *internal* to, so to speak, the realm of the two children. What I consider the prime merit of my analysis is that it conforms to, and helps to give a clearer expression of, this metaphysical idea.

In presenting my analysis, I shall concentrate on a numerical fact closely related to, if much more basic than, fact *II*:

IV. John and Carol *are two things*.

I argue that this fact must be analyzed in terms of

being two things (in short, two[2])

as its predicative part as follows:

1. Similarly, I think that it is a viable, if more controversial, view that "Neither John nor Carol is a dog" and "No man lives in the island" have the same overall syntactic structure.

2. Henceforth I shall abbreviate "being two things" and, less often, "to be two things" as "two" and "to be two", respectively.

> Fact *IV* consists in the instantiation of *being two things* by *John and Carol.*

It is this analysis that I encapsulate in the thesis that two is a property, i.e., a 1-place predicative component of facts. Surely, however, the analysis is applicable to a much broader range of facts, such as the following:

> John and Carol *sing together.*
> John and Carol *sit next only to each other.*
> John and Carol *are Jackie's only children.*

Thus I entertain the following, too, as properties needed to account for the structure of these facts:

> singing together;
> sitting next only to each other;
> being Jackie's only children.[3]

---

3. It seems that it is usual to consider singing together, for example, as a *relation* rather than as a *property*. But it is not a relation in the standard sense, i.e., the metaphysical counterpart of an *n*-adic predicate for a positive number *n*. For the following facts (Here let me assume that they are facts.), for example, contain it as a common predicative component:

> John and Carol *sing together.*
> John, Carol, and Jackie *sing together.*
> John, Carol, Jackie, and Chelsea *sing together.*

The natural view about the sentences that state these facts seems to be that they share the same *1-place* predicate whose argument place admits terms of various 'lengths': "John and Carol", "John, Carol, and Jackie", etc. My view of singing together is the metaphysical counterpart of this syntactical analysis: singing together is a 1-place relation (i.e., a *property*) that, to put it roughly, can be instantiated by John and Carol (taken together), or by John, Carol, and Jackie (taken together). Similarly, not-being-two-things (i.e., the 'complement' of two) and being-two-or-more-things (which is a property or relation closely related to

3.1.2 Irregularities of Numerical Facts

It is usual to think that extending the standard analysis to, e.g., fact *IV* is a non-starter. The grounds for this view have not been made explicit, but they seem to have much to do with the so-called irregularities of the fact. Consider the following fact, which involves the property of *being young*:

> *V.* John *and* Carol are young.

The following are (logical) consequences of this fact:

> John is young *and* Carol is young.
> *Something* is young.

But the counterparts of these two facts with respect to fact *IV*, the fact that John and Carol are two, are not its consequences; their 'opposites', namely, the following facts, hold:

> *It is not the case that* [John is two things *and* Carol is two things].
> *Nothing* (viz. no thing) is two things.

But this contrast between facts *IV* and *V* fails to support the conclusion that fact *IV* cannot be analyzed in terms of the property of being two (or that it is not a property that composes the fact). The contrast leaves open the possibility that being two is a property quite different from the property that composes fact *V* (i.e., being young):

> *Failure of Distribution:* Being two (or, in short, two) does not distribute over, e.g., John *and* Carol, in the way that being young does.

---

two) are properties, not relations. Thus even two, which might seem more amenable to treatment as a 2-place relation (e.g., being-not-identical-with), must be a 1-place relation (i.e., a property), which fails to be instantiated by Cicero, or Cicero and Tully (taken together), or John, Carol, and Jackie (taken together).

> *Failure of Generalization:* Fact *IV*, which consists in the instantiation of being two, does not have its *standard existential generalization* as a consequence.[4]

My analysis of fact *IV* is based on accepting these theses to account for its so-called irregularities. This contrasts it with the standard analysis, which can be seen to be based on taking the explanation that two is not a property at all. I shall argue in the next section that the standard analysis fails.[5] This gives a good reason for exploring the alternative explanation that the predicative component of fact *IV* is a property of quite a different kind.

3.1.3 Why Not the Set Analysis?

It is usual to reduce fact *IV* to a fact about the set {John, Carol}, which contains John and Carol (and only them) as members. On this view, fact *IV* is the same fact as the following:

---

4. I add "standard" to make it clear that what is meant here is only that the following is not a consequence of fact *IV*:

> *Something* is two things.

For the following is its consequence:

> *Some things* are two things (or, to put it more idiomatically, there are two things).

5. Someone who takes this view might consider reducing fact *IV* to the fact that the number of John *and* Carol is two. This fails to explain away the 'irregularities'. For this latter fact is also subject to the same phenomena: it is *not* the case that [the number of John is two *and* the number of Carol is two]; *Nothing* is such that the number of it is two.

VI. {John, Carol} is two-membered.

That is, the sentences used to state them, i.e., the following, are merely different expressions for the same fact:

(4) John and Carol are two things.
(6) {John, Carol} is two-membered.

But, the view holds, sentence (6) gives a good representation of the structure and components of the fact in question whereas sentence (4) states the fact in a misleading way: the fact *VI* that is stated by sentence (6) consists in the instantiation of a property (viz. *being two-membered*) by a set (viz. the set *{John, Carol}*); thus fact *IV* must also consist in the instantiation of the same property by the same object.[6]

I think that this analysis of fact *IV* (call it the *set analysis*) is wrong. Fact *IV* cannot be reduced to fact *VI* because fact *VI* is not a consequence of fact *IV*. To see this, consider the following facts:

VII. John is a thing, Carol is a thing, and John is not identical with Carol.
IV. John and Carol are two things.
VI. {John, Carol} is two-membered.
VIII. There is something (e.g., a set) that has at least one member.

6. How can this analysis deal with the 'irregularities' of fact *IV*? First, on this analysis, "and" does not occur in the correct statement of fact *IV*. (So the complex term "{John, Carol}" that occurs in sentence (6) must be paraphrased away without using "and". This can be done using the predicate "to be identical with John *or* Carol": "the only thing *x* (e.g., a set) such that for anything *y*, *y* is a member of *x* if and only if *x is identical with John or Carol*".) Second, fact *IV*, on this view, appears not to admit standard existential generalization only because being two things is mistakenly taken to be its predicative part; given that its genuine predicative part is the property of being two-membered, the fact that *something* (e.g., {John, Carol}) is two-membered can be seen to be its existential generalization.

I maintain three theses about these facts:

A. Fact *IV* is a consequence of fact *VII*.
B. Fact *VIII* is a consequence of fact *VI*.
C. Fact *VIII* is *not* a consequence of fact *VII*.

The desired conclusion that fact *VI* is *not* a consequence of fact *IV* follows directly from these three theses. Now, let me defend the theses. *Thesis A:* Granting that John is a thing and that Carol is a thing, they (namely, John and Carol) must be two different things *insofar as* John is not identical with Carol. *Thesis B:* I find this obvious and undeniable; no matter how one might analyze being two-membered, it must be a consequence of something's being two-membered that it has two members and so at least a member. *Thesis C:* Just because John is not identical with Carol (and both are things) does not mean that there is a set (or a set-like object) of which something is related as a member. Thus fact *VII* holds independently of fact *VIII*. This completes my argument against the set analysis (call it the *consequence argument*).[7]

---

7. The argument rests on the fact that Theses *A*-*C* get strong intuitive support. But the nature of the consequence relation is not well-understood. The modal relation of entailment that is defined in terms of *metaphysical* necessity is not an adequate analysis of the relation. Given the orthodox view on the modal status of sets (for this, see Kit Fine (1981)), fact *VII* modally entails fact *VIII*. This, however, gives no relief for the set analysis. For the fact that John is a child (i.e., fact *II*) and the fact that

II'. {John} has exactly one member that is a child

modally entail each other, which cannot be taken to mean that fact *II* consists in the instantiation of *having one member that is a child* by the set {John}. To do justice to the idea that *II* is not the same fact as *II'*, a distinction finer than the one available through modal entailment (which is defined in terms of metaphysical necessity) is needed. For more on this and related issues, see Fine (1994) and Joseph Almog (1991).

3.2 A Theory of Plural Properties

I present my account of fact *IV* in two stages. In this section, I elaborate on the analysis that decomposes the fact into the subject and predicative parts. In the next section, I give a further analysis of the fact by analyzing the predicative part.

3.2.1 The Predicative Part

I hold that the predicative part of fact *IV* is *two* (i.e., *being two things*). This property must be distinguished from, for example, being two-membered. A set can instantiate the latter property, but not the former. The set {John, Carol}, for example, is two-membered; it has two things, John and Carol, as members. But this does not make the set itself two things. In this respect, the set is no different from Jackie. She is only one thing, one human being, although she has two offspring (viz. John and Carol); the fact that two children have the offspring relation to Jackie does not make Jackie herself two things. It is the same with the set: the membership relation that relates exactly two things to the set makes the set itself two things no more than the offspring relation makes Jackie herself two children. Despite its relation to two things, the set itself is one, and only one, thing; it is only *its members* that are two.[8]

---

8. This is not to deny that the membership relation might in a sense tie the set more closely to John and Carol than the offspring relation ties Jackie to the children. There is, on my view, a good sense in which the membership relation is intrinsic to the set while the offspring relation is extrinsic to Jackie; John and Carol are essential to the being or existence of the set while they are not so with respect to Jackie. But such a difference in the nature of the relations makes no difference to the numerical status of the set and Jackie. A water molecule not only relates to three things (viz. two hydrogen atoms and an oxygen atom), but also consists of them. Still, the molecule itself is only one thing (viz. one molecule).

3.2.2 The Subject Part

The same consideration shows that the set analysis and my analysis differ with respect to the subject part of fact *IV* as well. What on my analysis instantiate two to constitute fact *IV* are: *John and Carol.*

These children, who are two things, must be sharply distinguished from the set {John, Carol}. Here I am not making the unchallenged point that neither John nor Carol is identical with the set. This is certainly correct, but deserves no special attention. My point is that John *and* Carol (taken together) are not the same as the set {John, Carol}. John *and* Carol are two things (viz. two children); the set is only one thing. We can see the contrast more clearly by comparing the following:

(a) John, Carol, and *the set {John, Carol}*;
(b) John, Carol, and *John and Carol.*

The former are three different things;[9] the latter are not three, but only two things. Thus the term "the set {John, Carol}" is not a stylistic variant of "John and Carol". The former term refers to something, an object, that differs from both John and Carol.

Underlying this distinction is a perfectly general point: Everything (i.e., every *thing*) is one, and only one, thing. It does not matter what kind of thing it is. Whether it is a child or adult, a human being or rabbit, material or immaterial, spatio-temporal or not, simple or complex, concrete (as a molecule is) or abstract (as a set is), and so on are all totally irrelevant. Similarly, it is irrelevant how many children, parts, or members it has. Insofar as it is a thing, it is only one thing. This general fact, together with the specific fact that sets, as considered in contemporary set theory, are things or objects, yields the sharp contrast that I draw between John and Carol, on the one hand, and the set thereof, on the other.

9. Accordingly, the set {John, Carol, {John, Carol}} is three-membered.

3.2.3 Instantiation

Now, I say that fact *IV* consists in the *instantiation* of two by John and Carol. This is not to say that two is instantiated by John *and also by* Carol. Neither child instantiates the property. John, who is a thing, is one thing, not two things, and it is the same with Carol. In general, nothing whatsoever instantiates two, because anything (i.e., any thing) is one thing only. But this does not mean that two is an uninstantiated property like the property of *being not identical with itself.* It is one thing to say that it is not instantiated by any one thing, quite another to say that it is not instantiated at all. Two is instantiated by John *and* Carol, and so by some things (e.g., children).

Multiple Instantiation: There is, I think, a sense in which *being a child* is instantiated by more than one thing, a large number of things. But it is not in this sense that I say that two is instantiated by some thing*s*. Being a child is instantiated by John *and also by* Carol. One might put this by saying that both John and Carol instantiate the property. So one might perhaps say that some things, more than one thing, instantiate it. This helps to distinguish being a child from, for example, *being a child of Hillary*, a property that is not instantiated by more than one thing. These properties are distinguished from each other in that the former is, as I shall say, *multiply instantiated* while the latter is not. In this respect, two is akin to the former property, not the latter; it is multiply instantiated, instantiated by (a) John and Carol, *and also by* (b) Hillary and Chelsea. But it is not to point out this characteristic of the property that I say that it is instantiated by some things.

Plural Instantiation: Two is instantiated by John and Carol. This is a characteristic that two shares with, for example, *being Jackie's only children*, but not with being a child. Consider the property of being Jackie's only children. This property is not multiply instantiated. It is instantiated by John and Carol, but by no other thing or things; it is not instantiated by John, nor by Carol, nor by anything else, nor by any things other than John and Carol. Nevertheless, there is a clear sense in which it is instantiated by some things, the very sense in which it is instantiated by John *and* Carol. Because the children (taken together) instantiate two, there are *some things* (viz. those children) that instantiate it. Clearly, two is instantiated by John and Carol, just as being Jackie's only children is instantiated by them. Accordingly, it is instantiated by some things in the same sense in which this property is so instantiated. To distinguish this idea from the idea of multiple instantiation, it seems to help to give a cue, so to speak, by inserting the parenthetical remark "taken together" (or "collectively") as I have done. Obviously, however, such phrases cannot be taken literally; they must be considered as cues only. It sometimes helps to talk of "river-bank" to convey the idea meant, although the so-called money-banks can be located alongside a river. Likewise, it might usually help to talk of "taken together" or "collectively", but not always. One might be lead to think that the activity of collect*ing* (or tak*ing*) John and Carol together (or at least, their capacity or potential for being collected) is somehow tied to their being two things. Needless to say, however, not even a shadow of collecting or taking together is needed for some things to be two. John and Carol's being two things in no way depends on the chance that there is a sentient being that is capable of 'collecting' (or conceiving) them together. Nor does it depend on the existence of something (e.g., the set {John, Carol}) the nature of which consists in 'comprehending' them. In order to dissociate two from any association with collecting, taking together, comprehending, and so on, let me introduce the neutral term "*plural* instantiation". Say that a property is *plurally instantiated by* some things (e.g., John and Carol), if it is instantiated by them, so to speak, taken together; accordingly, say that a property is *plural* (or *plurally instantiated*), if it is plurally instantiated by some things.[10]

10. Accordingly, say that a property is *singular* (or *singularly instantiated*) if it is instantiated but not plurally. In drawing these distinctions, I do not entertain different kinds or species of instantiation but simply give classifications of properties into different categories in terms of the *same* relation of instantiation. This is comparable to distinguishing, e.g., sets into different groups (the empty set, singletons, and other multi-membered sets) in terms of the same relation of membership. It is not that being a child has a different kind of relation to a particular child (e.g., John) depending on how many children there happen to be in the world. Similarly, on my view, it is not that different species of the generic relation of instantiation relate the property of *being all the children living in White House* to those who instantiates it (at a time) depending on how many children live there (at that time).

3.2.4 Various Kinds of Plural Properties

Now, it should be clear that two shares the characteristic of plural instantiation with the following properties:

> singing together;
> sitting next only to each other;
> being Jackie's only children.

These, like two, are plural properties that compose the following facts, respectively:

> John and Carol sing together.
> John and Carol sit next only to each other.
> John and Carol are Jackie's only children.

The properties, including two, are plural properties of a special sort. Plural properties of another sort include:

> being one or more things;
> being a child or some children of a congressman.

These properties do not have the so-called irregularities, failures of distribution and (standard) generalization.[11] But once they are seen to be plural properties, their 'regularities' (viz. affinity with multiply instantiated singular properties) arise as unusual features for plural properties. In any case, once we have seen the need for the broader framework that includes plural as well as singular properties, we can use the so-called irregularities as the basis for isolating and describing a subspecies of properties under the general category of plural properties.

Still, focusing on, e.g., two, which turns out to be a plural property of a special sort, is critical to establishing the need for such a framework. Another

---

11. For this reason, it is easy to confuse them with the corresponding singular properties.

characteristic specific to two has also been useful to achieving this aim. Two is an *intrinsic* property. Not every plural property is an intrinsic feature of some things that plurally instantiate it. Being Jackie's only children, for example, is not such a property. John and Carol (plurally) instantiate the property, but only on account of their relation to someone else, viz. Jackie. Thus their instantiation of the property is not a matter *internal* to the realm of those children.[12] It is the same with the plural property of, for example, *being all the members of something that is two-membered*. By contrast, two is an intrinsic feature of John and Carol (as 'taken together'). Their plural instantiation of the property is a matter internal to the realm of the children. Thus the holding of fact *IV*, unlike that of the fact that *John and Carol are all the members of something that is two-membered*, does not depend on the existence of an object lying outside the realm. We can see that this characteristic of two is useful in arguing against the set analysis of fact *IV*, an analysis that can be seen to result from applying a general scheme of reducing a 'plural' fact (e.g., fact *IV*) to a 'singular' fact (e.g., fact *VI*). The idea behind the consequence argument is this: fact *IV*, being a fact about John and Carol only, is internal to the realm of the children; but fact *VI* is not so. Now, to see that fact *IV* is *exclusively* about John and Carol, it is useful to note that two is an intrinsic property. In the next section, I give an analysis of two that confirms this view about the property.

3.3 Two As an Intrinsic Plural Property

Consider the following condition:

> There is something x and there is something y such that — satisfy the following clauses:

12. So the property is subject to further analysis in terms of the (plural) relation *being the only children of* and the object *Jackie*. Thus the fact that John and Carol are Jackie's only children is as much a fact about Jackie as one about John and Carol.

(*i*) x is one of —, and y is one of —;
(*ii*) x is not identical with y;
(*iii*) Anything that is one of — is identical with x or y.

It is straightforward to see that any two things (i.e., any things that are two) satisfy the condition, and *vice versa.*

John and Carol, for example, who are two things, satisfy the condition.[13] For, notice, all of the following hold:

(*i'*) John is a thing that is one of *John and Carol*, and Carol is a thing that is one of *John and Carol.*
(*ii'*) John is not identical with Carol.
(*iii'*) anything that is one of *John and Carol* is identical with John or Carol.

Thus, by assigning John as the value of the variable "x" and Carol as that of "y" in the condition, we can satisfy all the three clauses, (*i*)-(*iii*), with the argument place (marked by "—") in the clauses filled in with "John and Carol". By contrast, as we can easily see, *John*, who is not two things but only one thing, fails to satisfy the condition; in this case, no assignment can be made to the variables "x" and "y" so that (*i*) and (*ii*) are both satisfied.[14] *John, Carol, and Jackie*, who are not two but three things, also fail to satisfy it; (*i*) and (*iii*) cannot be jointly satisfied. In general, (*i*) and (*ii*) can be jointly satisfied if the things in

---

13. That is, the following holds with *John and Carol* (taken together) assigned as the value of "they" (or "them"):

> There is something x and there is something y such that *they* satisfy the following:
>
> (*i*) x is one of *them*, and y is one of *them.*
> (*ii*) x is not identical with y.
> (*iii*) Anything that is one of *them* is identical with x or y.

14. Similarly, *Cicero and Tully*, who are one and the same human, fail to satisfy it.

question are *at least two*; (*ii*) and (*iii*), if they are *at most two* things. Thus, all the three clauses can be jointly satisfied if, and only if, the things in question are exactly two things.

Now, such an agreement by itself is neutral to the nature of property two. For example, even those who grant that God exists and loves everything that is identical with itself (and further that both are necessarily so) might still hold that *being loved by God* and *self-identity* are different properties (for, they might argue, the latter is an intrinsic property of John while the former is not). This situation contrasts with that with the two properties of *being an uncaused event* and *being an event of which nothing is a cause.*[15] These properties seem to have more substantial relation to each other. They, it seems, are the same property; the former property must be analyzed in terms of the relation of *being a cause of.* I think that it is the same with *being two* and *satisfying the above condition*; that is, on my view, they are the same property, and the following is a correct analysis of being two:

> *Analysis of Two*: For (something or) some things *to be two things* is for them to be such that there is something x and there is something y such that they satisfy the following:
>
> (*i*) x is one of them, and y is one of them.
> (*ii*) x is not identical with y.
> (*iii*) anything that is one of them is identical with x or y.

This completes my analysis of two. Let me now explain that this analysis conforms to and clarifies the idea that two is an intrinsic property of, e.g., John and Carol.

Notice that the condition that expresses the analysans of the property can be equally formulated as follows:

> There is something x *that is one of* — and there is something y *that is one of* — such that — satisfy the following clauses:

15. For a more substantial example, which is also more controversial, consider the pair of *being a water molecule* and *being a molecule consisting of two hydrogen atoms and one oxygen atom in such-and-such a way.*

(a) x is not identical with y.
(b) Anything *that is one of* — is identical with x or y.

In the condition thus reformulated, every quantifier occurs *bounded* by the clause "that is one of —". Consequently, whether or not some (one or more) things satisfy the condition is a matter entirely *internal to* the realm of the things in question. It is totally *independent of* what it is like outside the realm; in particular, whether or not there is *anything that is* not *one of* them is irrelevant to their satisfaction (or non-satisfaction) of the condition. This makes precise the idea that fact *IV*, for example, is a fact about John and Carol *only*. Thus, according to my account of the fact, whether there is anything other than the children (e.g., the set {John, Carol}) or how it is with that thing is irrelevant to the instantiation of two by John and Carol.

3.4 Concluding Remarks

In this chapter, I have given a case for plural properties. I have argued that the broader framework that accepts the existence of plural, as well as singular, properties is coherent, legitimate, and natural. I have also argued that the numerical fact *IV*, for example, cannot be adequately accounted for without accepting two as a plural property. The set analysis of the fact conflicts with strong intuitions on what the fact is about. My analysis of the fact, by contrast, conforms to the intuitions and helps to clarify and explain them. Thus the broader framework is irreducible and fruitful as well.

4. What Numbers Should Be

What is the number one? Gottlob Frege raises this question in *The Foundations of Arithmetic*.[1] He gives the answer that it is the extension of a certain concept.[2] In general, on his account, natural numbers (such as one, two, or three)[3] are "self-subsistent objects" of a certain kind, extensions of concepts. I propose a radically different account in this chapter. On my account, natural numbers are properties: the number one is the property of *being one thing*, the number two is the property of *being two things*, and so on. To defend this account, I shall present a general scheme of analyzing apparently diverse numerical facts, such as the following:

I. *Two* plus one is three (i.e., 2+1=3).

II. Bill and Hillary are *two* Americans.

The account meshes well with a uniform analysis of these and other numerical facts. By contrast, Frege's or other prominent accounts of number rest on inadequate analyses of them.

---

1. Frege (1984).

2. That is, ⌜equal to the concept ⌜identical with the number zero⌝⌝, where the number zero, on his view, is the extension of the concept ⌜equal to the concept ⌜not identical with itself⌝⌝ (here corners are used to make terms designating Fregean concepts).

3. The discussion in this chapter is limited to the natural numbers (greater than zero). So I shall henceforth talk of "number" to mean natural numbers. Most of my views on natural numbers apply to their infinite 'cousins' as well, but I leave it for another occasion to extend the views to infinite numbers.

In section 4.1, I examine the Fregean conception that numbers are a kind of objects and argue that the conception leads to an incorrect analysis of ordinary numerical facts, such as *II*. In sections 4.2 and 4.3, I present my analysis of various numerical facts. In section 4.2, I argue that fact *II*, for example, must be analyzed in terms of a property instantiated by Bill and Hillary (taken together), i.e., the property of *being two Americans*. In section 4.3, I extend the analysis of fact *II* to other numerical facts, especially arithmetical facts, such as *I*. Finally, in section 4.4, I argue that the analysis supports my account of number because, on the analysis, numerical properties, such as *being one thing* or *being two things*, are common components of various numerical facts.

4.1 Are Numbers Objects?

I think that an account of number is best seen as a part or continuation of, or perhaps an annex to, a general scheme of analyzing various numerical facts. It is usual to present and support accounts of number on the basis of schemes of analyzing some, if not all, kinds of numerical facts–or else the true sentences used to state them. So in order to understand and assess the accounts, it is necessary to consider the schemes that they rest on. The Fregean conception that numbers are objects of a certain kind is based on a natural scheme of analyzing arithmetical facts, such as *I*. The thought underlying the conception, very briefly, is this. Consider fact *I* in light of the true sentence often used to state it:

(1) *Two* plus one is three.[4]

The numerical terms (e.g., "two"), it seems, occur as singular terms in this sentence. Because there is the presumption that the canonical function of a singular term (e.g., "Bill") is to refer to an object, of one kind or another, it is natural to think that the sentence is true because the singular term "two", for example, refers to an object. If so, one might conclude, the number two, which is a component of fact *I* just as the term "two" is a component of sentence (1), must be the object the term refers to. But the Fregean conception of numbers as

4. Or alternatively, "2+1=3."

objects meets serious difficulties in accommodating ordinary numerical facts, such as *II*. Consider fact *II* in light of the sentence used to state it:

(2) Bill and Hillary are *two* Americans.

It is far from natural to think that an object, one identified as the number two, is a constituent of fact *II*; in sentence (2), the numerical term "two" is not used as a singular term. Moreover, no object except Bill and Hillary seems to constitute the fact stated by sentence (2). To their being two different Americans, it seems, it is totally irrelevant whether there is another object that can be identified as a number or not. Clarifying and consolidating this thought about fact *II* leads to my argument against the Fregean conception.

It is convenient to present the argument with respect to a specific account of number that falls under the conception, i.e., an account that identifies some specific objects as numbers. Consider the account that identifies numbers with sets as follows:[5]

One = the set $\{x|\ x$ is equinumerous with $\{\varnothing\}\}$
Two = the set $\{x|\ x$ is equinumerous with $\{\varnothing, \{\varnothing\}\}\}$
Three = the set $\{x|\ x$ is equinumerous with $\{\varnothing, \{\varnothing\}, \{\varnothing, \{\varnothing\}\}\}$
. . .

This account is based on a scheme of analyzing arithmetical facts that renders fact *I* (i.e., the fact that 2+1=3), for example, to the following:

III. The union of any member of the *set-number* 2 and any disjoint member of the *set-number* 1 is a member of the *set-number* 3 (and *vice versa*),

where "the set-number 1", for example, abbreviates "$\{x|\ x$ is equinumerous with $\{\varnothing\}\}$". Extending the scheme to one that applies to ordinary numerical facts as well leads to the following analysis of fact *II* (i.e., the fact that Bill and Hillary are two Americans):

5. Here "$\varnothing$" refers to the empty set. This account is like Frege's account except that his account identifies numbers as extensions of concepts rather than sets.

IV. The set {Bill, Hillary} is a member of the set-number 2, and Bill is an American and Hillary is an American.

But this, I argue, is not a correct analysis of fact *II*.

To see this, compare the following facts:

V. Bill is not identical with Hillary, and Bill is an American and Hillary is an American.
II. Bill and Hillary are two Americans.
VI. There is something (e.g., the set-number 2) that has at least one member.

Given that both Bill and Hillary are Americans, they must be two different Americans *insofar as* Bill is not identical with Hillary. So,

Thesis A: Fact *II* is a consequence of fact *V*.

By contrast,

Thesis B: Fact *VI* is *not* a consequence of fact *V*.

For just because Bill is not Hillary (and they are both Americans) does not mean that there must be a set (e.g., the set-number 2) that contains something as a member. Given these theses, it follows that fact *VI* is *not* a consequence of fact *II*. But fact *VI is* a consequence of fact *IV*. Thus, fact *IV* is not a correct analysis of fact *II*. This completes my argument that the account that identifies numbers (e.g., two) with some sets (e.g., the set-number 2) leads to a wrong analysis of ordinary numerical facts.

Notice that the argument applies *mutatis mutandis* to other accounts of number that fall under the Fregean conception. Such an account consists in identifying numbers (e.g., two) as other specific objects (e.g., the set {∅, {∅}}). The account leads to an analysis of fact *II* in terms of the object identified as the number two (and its relation to Bill and Hillary). Thus we can show that the alleged analysans of fact *II* is not a consequence of fact *II*—the analysans is not a consequence of fact *V*. The account that identifies the number two with the set

{∅, {∅}}, for example, leads to an account that identifies fact *II* with the following:

> The set {Bill, Hillary} is equinumerous with {∅, {∅}}, and Bill is an American and Hillary is an American.

This fact, which requires the existence of the set {∅, {∅}}, is no more a consequence of fact *V* than fact *IV* is. Thus this account, too, leads to a wrong analysis of fact *II*.

Those who hold the Fregean conception of number might reply that the question of how to analyze ordinary numerical facts, such as *II*, is irrelevant to the issue of what numbers are. But those who do so to defend the conception cannot take the same stance on analysis of arithmetical facts, such as *I*; the main ground for the conception is that it is based on a natural analysis of these facts. Can they justify the double stance on, e.g., facts *I* and *II*?

Some might appeal to different ways in which the word "two" is used in the sentences used to state the facts. It is used as a noun in sentence (1), but as an adjective, it seems, in sentence (2). So they might argue that a sharp distinction must be drawn between *the-two-in-fact-I* and *the-two-in-fact-II*, i.e., the components of the facts that are associated with "two" in sentences (1) and (2), respectively. Then they might maintain that the Fregean conception is correct since it only concerns the arithmetical numbers, such as the-two-in-fact-*I*.

But this is not a viable defense of the conception. It ignores the close connection between the two kinds of numerical facts. According to the view that there is no connection between the-two-in-fact-*I* and the-two-in-fact-*II*, it would be a fallacy to appeal to fact *II* as in the following argument:

> Bill and Hillary are *two* Americans. Chelsea is another American. So the former and the latter (i.e., Bill and Hillary, and Chelsea) are three Americans; for *two* plus one is three (i.e., 2+1=3).

To the contrary, the argument should strike one as unexceptionable. There is nothing odd about the appeal to the arithmetical fact in connection with the ordinary numerical facts stated therein. If so, the apparent affinity between the arithmetical and ordinary numerical facts cannot be a mere linguistic artifact, the result of the equivocal uses of the English word "two". I do not deny that the word "two" is used differently in sentences (1) and (2), but I reject the idea that its different uses can be assimilated to the different uses of, e.g., "bank". The uses of this word have no connection to each other as we can see clearly in, for example, the following argument, which is fallacious because it contains different uses of "bank":

> Bill walked on a $bank_1$. So he walked on a financial institution; a $bank_2$ is a financial institution.

But it is not so with the two different uses of "two" in, for example, (1) and (2). Despite their difference in grammatical character, they are so closely related as to support logical connections between the two kinds of numerical facts (or their statements).

An adequate scheme of analyzing numerical facts, we have seen, must apply to both arithmetical and ordinary numerical facts and help to explain their logical connections. My aim in the following sections is to present an account of number based on such a scheme.

4.2 Plural Properties as Components of Numerical Facts

In this section, I present my analysis of, for example, fact *II*:

> *II.* Bill and Hillary are two Americans.

I analyze this fact into two components:

> The subject part: *Bill and Hillary*;
> The predicative part: *being two Americans.*

On the analysis, the fact consists in the instantiation of a property, viz. *being two Americans*, by two objects, viz. *Bill and Hillary* (taken together). This analysis is a generalization of the standard analysis of, for example, the following:

> *VII.* Bill is an American.

On the standard analysis, this fact is composed of two components:

> The subject part: *Bill*;
> The predicative part: *being an American*.

The fact, on the analysis, consists in the instantiation of the predicative part by the subject part. Similarly, I maintain that fact *II* must be analyzed in terms of predicative and subject parts and that the predicative part (viz. being two Americans) is instantiated by the subject part (viz. Bill *and* Hillary (taken together)) to form the fact.

Notice that I add the phrase "taken together" in parenthesis to warn against confusing my analysis with that on which fact *II* consists in the instantiation of its predicative part both by Bill and by Hillary. This is clearly a wrong analysis. Bill does not instantiate being *two* Americans; he is not *two* Americans, but only one such. And it is the same with Hillary. It is Bill *and* Hillary, not individually but as taken together, that are two Americans. The phrase "taken together" helps to clarify this. But it cannot be taken literally. The disambiguation device, like any other device that I know of, has misleading connotations. It suggests the idea that Bill and Hillary are *two* Americans if (and only if) they are taken together. This is a wrong idea.[6] Whether taken together (by someone) or not, Bill and Hillary are *two* Americans; they are two Americans insofar as both of the following hold:

> (i) Bill is not identical with Hillary.
> (ii) Bill is an American and Hillary is an American.[7]

---

6. This is why I put the phrase in parenthesis.
7. On my analysis, fact *II* is logically equivalent with the following fact:

> Bill is not identical with Hillary, and Bill is an American and Hillary is an American.

Yet, as I argue below, fact *II* cannot be identified or analyzed as this latter fact.

Thus their being two Americans is independent of even the possibility of their being collected, physically or mentally, by anyone or anything, sentient or not.

In particular, they are two Americans regardless of the existence of some one thing that in a sense comprehends them, such as the set {Bill, Hillary}. Accordingly, there is a sharp distinction between them, on the one hand, and the set {Bill, Hillary}, on the other. The humans are two Americans; the set is *not*. The set in a sense *consists of* two American adults, but this no more makes the set itself two American adults than Chelsea's being a child of two American adults makes Chelsea herself two American adults. Consequently, my analysis of fact *II* must be clearly distinguished from that on which the fact consists in the instantiation of *being two Americans* by *the set {Bill, Hillary}*. This analysis is flagrantly wrong. The set fails to instantiate the predicative component; it is not two Americans. It is also wrong to identify my analysis with that on which fact *II* consists in the instantiation of *being two-membered and having only Americans as members* by the set. Bill and Hillary (taken together), we have seen, must be distinguished from the set. Being two Americans must also be distinguished from being two-membered etc.; the set instantiates the latter property (it is two-membered), but not the former (it is not two Americans). And, on the analysis, it would be essential to Bill and Hillary's being two Americans that there is something (viz. the set) that comprehends them.

Notice also that, on my analysis, the predicative component of fact *II*, i.e., *being two Americans*, is a property, not a relation. Thus the analysis contrasts with that on which the fact consists in Bill's bearing a relation to Hillary. Let me explain.

It is usual to distinguish two kinds of *predicative components* of facts: *properties* and *relations*. I think the related notions of property and relation can be understood in the context of analyzing, for example, the following facts:

VII. Bill is an American.
VIII. Bill loves Hillary.

On the standard analysis of these facts, the property of being an American and the relation of loving are the predicative components of facts *VII* and *VIII*, respectively: fact *VII* consists in the instantiation of the property by Bill; fact *VIII*, in Bill's bearing the relation to Hillary. Now, what distinguishes properties from relations is how many argument places they have: a property has only one argument place; a relation, two or more. On my analysis, then, the predicative component of fact *II* has only one argument place. To see that this is correct, consider the following facts:

Bill *is not two Americans.*
Bill and Boris *are not two Americans.*
Bill, Hillary, and Chelsea *are not two Americans.*
The children in the Common Room *are not two Americans.*

These facts, it seems, have a common predicative component: *not being two Americans.* If so, the common predicative component must have only one argument place, one capable of, so to speak, admitting (or accommodating): Bill; or Bill and Boris (taken together); or Bill, Hillary, and Chelsea (taken together); etc. Now, *not being two Americans* is the *complement* of the predicative component of the fact that Bill and Hillary are two Americans: *being two Americans.* Consequently, this must also have only one argument place. That is, it must be a property.[8]

---

8. And the argument place must have the same logical character as that of its complement; it must also be capable of admitting any number of things, namely, anything (e.g., Bill) or any thing*s* (e.g., Bill, Hillary, and Chelsea (taken together)). It is useful to note the parallel that my analysis of fact *II* draws to the natural analysis of sentence (2), which states the fact. The sentence, it seems, is composed of the grammatical subject "Bill and Hillary" and the predicate "to be two Americans". What my analysis takes to be the subject and predicative parts of the fact are the metaphysical counterparts of these grammatical components. Given the grammatical analysis, it is natural to say that (2) is true because the predicate "to be two Americans" is true of Bill *and* Hillary, thought not of either one of them. Similarly, on my analysis, fact *II* consists in the instantiation of the predicative part by those people (though not by either of them). Now, compare (2) with the following sentences:

Bill *is two Americans.*
Bill, Hillary, and Chelsea *are two Americans.*
The children in the Common Room *are two Americans.*
They *are two Americans.*

4.3 Analysis of Numerical Facts

The above analysis of fact *II* applies, *mutatis mutandis*, to the following as well:

> *IX*. Bill and Hillary *are two things*.[9]

So I analyze fact *IX* into two components: (a) *Bill and Hillary*; and (b) *being two things*. Accordingly, its predicative component, being two things, is a property, not a relation.

Now, notice that being two things is a property that underlies being two Americans. The latter property can be analyzed in terms of the former as follows:

> For some things (viz. something or some things) *to be two Americans* is for them to be such that (i) they *are two things* and that (ii) they are Americans.

This analysis of the property that forms fact *II* yields a further analysis of the fact:

They seem to have a common component with (2), namely, the predicate "to be two Americans" (whose singular and plural forms are "is two Americans" and "are two Americans", respectively). The predicate combines either with singular terms (e.g., "Bill") or with plural terms (e.g., "Bill and Hillary") to form grammatical sentences; that is, it is a *1-place* predicate whose argument place is neutral with respect to singular and plural terms. My view that being two Americans is a property is the metaphysical analogue of this view on the predicate.

9. It has also applications to, for example, the following facts:

> Bill and Hillary *admire each other*.
> Bill and Hillary *sing together*.
> Bill and Hillary *are Chelsea's only parents*.

> Bill and Hillary *are such that they are two things and that they are Americans.*[10]

Thus the numerical property of being two things is contained in fact *II*–just as *being a male*, a property that occurs in the analysis of being a bachelor, is contained in the fact that John is a bachelor.

Now, let us consider arithmetical facts, such as fact *I*:

> *I.* Two plus one is three (i.e., 2+1=3).

I propose to analyze this fact in terms of numerical properties:

> If some things (viz. something or some things) *are two things* and if something *is one thing* that is not one of the former, then the former and the latter *are three things* (and *vice versa*[11]).

To put it roughly, any two things together with another thing are three things (and *vice versa*). This analysis of fact *I* yields a straightforward account of the connection between arithmetical and ordinary numerical facts. Consider, for example, the following:

> *X.* If Bill and Hillary *are two things* and if Chelsea *is one thing* that is not one of the former, then the former and the latter *are three things*.

Fact *I* (or, more precisely, its analysans) is, so to speak, a universal generalization of this fact and thus yields it as a logical consequence.

Notice that this connection justifies the appeal to the arithmetical fact in, e.g., the following argument:

---

10. Notice that in the sentence used to state the analysis, the italicized complex predicate is used to indicate a complex property.

11. That is: if some things are three things, then there is some one thing such that the former excepting the latter are two things.

> Bill and Hillary are two things. And Chelsea is another thing. Therefore, the former and the latter are three things; for two plus one is three.

Similarly, it is straightforward to justify the appeal to the fact in the argument discussed in section 4.1:

> Bill and Hillary are two Americans. And Chelsea is another American. Therefore, the former and the latter are three Americans; for two plus one is three.

To do so, we need to show that the following fact is a consequence of fact *I* (or its analysans):

> *XI.* If Bill and Hillary *are two Americans* and if Chelsea *is one American* that is not one of the former, then the former and the latter *are three Americans.*

This fact is a consequence of the analysans together with the following fact:

> If Bill and Hillary are Americans and if Chelsea is an American, then the former and the latter are Americans.

And, clearly, this fact holds logically. So fact *XI* is a consequence of the analysans of fact *I*. I think the above account of logical connections between the two kinds of numerical facts and the consequent justification of the intuitive appeal to the arithmetical fact to support an argument on ordinary numerical facts renders strong support to my analysis of the arithmetical fact *I*.

Let me complete this section with an analysis of the numerical properties that, on my view, constitute various numerical facts. Here is my analysis of the property of being two:

> For some things (viz. something or some things) *to be two things* is for them to be such that there is something *x* and there is something *y* such that the following hold:
>
> (*i*) *x* is one of them, and *y* is one of them.

(*ii*) x is not identical with y.
(*iii*) Anything that is one of them is identical with either x or y.

It is straightforward to give a similar analysis of other numerical properties, such as being one thing or being three things, that correspond to the natural numbers (greater than zero).[12] The analysis of the numerical properties yields my ultimate analysis of the various numerical facts.[13] The fact that Bill and Hillary *are two things* (i.e., fact *IX*), for example, can be reduced to the following:

> *Bill and Hillary* are such that there is something x and there is something y such that the following hold:

(*i*) x is one of them, and y is one of them.
(*ii*) x is not identical with y.
(*iii*) Anything that is one of them is identical with either x or y.

We can use this analysis to confirm that fact *IX* holds. Bill (assigned as the value of "x") and Hillary (assigned as the value of "y") satisfy conditions (*i*)-(*iii*); that is to say, all the following facts hold:

(*i'*) *Bill* is one of them [i.e., Bill and Hillary], and *Hillary* is one of them.
(*ii'*) *Bill* is not identical with *Hillary*.

---

12. Or they can be recursively characterized in terms of their predecessors as follows:

> For some things (viz. something or some things) to *be* $(n+1)$*-many things* is for them to be such that there is something x such that the following hold:
> (*i*) x is one of them.
> (*ii*) They except x *are* n*-many things*.

13. We can use the analysis to confirm, e.g., the following facts as well:

> Bill *is* not *two things*.
> Bill, Hillary, and Chelsea *are* not *two things*.

(*iii′*) Anything that is one of them is identical with either *Bill* or *Hillary*.[14]

4.4 Numbers Are Properties

On my analysis, the property of being two things is a common component of, for example, the following facts:

I. Two plus one is three.
II. Bill and Hillary are two Americans.

Similarly, the properties of being one thing, being three things, and so on are also common components of such facts as the following:

---

14. Furthermore, the analysis helps to clarify the truth of Thesis *A*: fact *II* (i.e., that Bill and Hillary are two Americans) is a consequence of fact *V* (i.e., that [Bill is not identical with Hillary, and Bill is an American and Hillary is an American]). To see this, we need to confirm that facts (*i′*) and (*iii′*), where "them" means Bill *and* Hillary (taken together), are consequences of the fact that each of Bill and Hillary is a thing (e.g., an American). This is a generalization of the principle that Bill's being the thing identical with himself is a direct consequence of his being a thing. Insofar as *Bill* is a thing, he is precisely the thing that satisfies the following condition:

— is identical with Bill.

Hence, the latter principle. Similarly, I think, insofar as each one of *Bill and Hillary* is a thing, they are precisely the things that satisfy the following conditions, which imply (*i*) and (*iii*):

Anything that is identical with either Bill or Hillary is one of —.
Anything that is one of — is identical with either Bill or Hillary.

Thus facts (*i′*) and (*iii′*) are direct consequences of the fact that [Bill is a thing, and Hillary is a thing].

Three plus one is four.
Chelsea is one, and only one, child.
Bill, John, and Boris are three politicians.

The numerical character of the apparently diverse facts and the logical relations among them rest on the presence of those properties in them. Moreover, there are no other components of those facts that have a comparable standing; in particular, no object can be seen to be a common component of facts *I* and *II*. So it is natural to identify natural numbers as those numerical properties as follows:

The number *one* is the property of *being one thing*.
The number *two* is the property of *being two things*.
The number *three* is the property of *being three things*.
. . . .

Thus, I conclude, natural numbers are properties.

5. How Sets Are Determined by Their Members

What is there? Everything, a famous philosopher has once said. What is included in the range of things? Tom and Julie, children who can sing together, are denizens of the world. And if current mathematics is to be trusted, some of the denizens are: sets. There are, for example, singleton Tom, singleton Julie, and the doubleton that consists of the two children. How are Tom and Julie related to the three sets? They are sets directly based on the children. Call things like Tom or Julie *mundane objects*, and sets that consist only of so-and-so's ***sets** of so-and-so's*. The idea is then: Sets of mundane objects are based on mundane objects. Further, the population of the sets is determined by that of mundane objects. Granting that Tom and Julie exist as mundane objects, the above-mentioned sets exist as sets of them; if in addition the children are the only mundane objects, the three sets are the only sets thereof.[1] In this way, the population of nothing but mundane objects affects that of the sets thereof. It is

1. The empty set is perhaps an exception, if it can be said to be a set *of mundane objects* in the degenerate sense. An *empty set* is a set with no member; a *singleton*, a set with only one member; a *doubleton*, a set with exactly two members. *The axiom of extensionality* states that sets having the same members are the same. (So there is only one empty set, *the* empty set.) $\{a_1, \ldots, a_n\}$ is the (unique) set that has exactly $a_1, \ldots, a_n$ as members; singleton a is the set $\{a\}$.

this idea of determination that I want to make clear in addressing the following question:

(Q) What sets of mundane objects are there?

The following answer, I argue, has the merit of conforming to and clarifying the idea:

(N) *Any number of* mundane objects form a set (viz. a set thereof).

This feature contrasts (N) with the standard answer to question (Q):

(C) *Any collection of* mundane objects is a set (viz. a set thereof).

(N) appears to be similar to (C). The critical difference lies in the two italicized phrases.

5.1 The Hierarchy of Sets and the Determination of Sets by Their Members

In "What is Cantor's Continuum Hypothesis?", Kurt Gödel has remarked that "as far as sets occur in mathematics . . . they are sets of integers . . . or sets of sets of integers, or etc." (1947/64, 474). This statement indicates the view that sets occur in what is called the *simple hierarchy*. On this view, there are some underlying objects that are not sets (namely, what I call *mundane objects*[2]), and sets directly based on the underlying objects, and then sets directly based on these sets, and so on. We can say that they occur at the basic level, the 1st level, the 2nd level, and so on, respectively. Suppose, merely for the sake of simplicity, that the mundane objects are the integers. Then things occur at the levels as follows:

---

2. They are usually called *urelements* or *individuals*.

basic level: integers
the 1st level: sets of integers
the 2nd level: sets of *sets of integers*
. . . .

In this way, on the view, every set occurs at one level or another in the hierarchy of levels.

The standard account of set is shaped, by and large, by the account that Gödel has presented as an improvement on the above account.[3] On the standard account, sets occur in the so-called *cumulative* hierarchy. The levels of the cumulative hierarchy in which the mundane objects are, for example, the integers contain the following objects:

basic level: integers
the 1st level: sets of integers
the 2nd level: sets of *both integers and sets of integers*
. . . .

The difference between the simple and the cumulative hierarchies first shows up at the second level. The mixed set {1, {1}}, for example, occurs at the second level of the cumulative hierarchy (its members are *either* integers *or* sets thereof), but not of the simple hierarchy.

Now, the above-mentioned idea of population determination can be stated in hierarchical terms as follows:

> The population of the first level in the hierarchy is determined by that of the basic level.

A generalization of this idea is another major component of the standard account of set:

---

3. Gödel (1947/64, 474f). See also Boolos (1971), Drake (1974, Chapter 1), Shoenfield (1977), and Moschovakis (1980, 448-51).

> The population of a given level in the hierarchy is determined by those of the levels below it.

To put it in hierarchical terms, my aim is to clarify the way the population of each level is determined by those of lower levels. To achieve the aim, I begin by looking for an answer to the following question:

> (Q1) What sets occur at each level, granting that the so-and-so's occur at levels below it?

This is an important question to ask for those who, assuming that sets occur in a hierarchy, address the ontological question "What is there and, in particular, *what sets are there*?" In raising (Q1), however, I aim at more than helping to answer this question. Any correct and informative answer to (Q1) might be satisfactory as far as the ontological question is concerned, but such answers can still be unsatisfactory because they fail to conform to, and clarify, the idea of population determination. As I shall explain in the next section, the standard answer to (Q1), which generalizes the answer (C) to question (Q), is an answer of this kind.

I shall henceforth focus on a question that is a special case of (Q1):

> (Q2) What sets occur at the *first level* of the hierarchy, granting that the so-and-so's are the things occurring at the basic level?

To be sure, I demand an answer more informative than the trivial answer that it is *the sets of so-and-so's* that occur at that level. The following formulation, meant to be a shorthand for (Q2), helps to forestall this answer:

> (Q) What sets of mundane objects are there?

And this formulation has the additional merit of succinctness, which comes from eliminating talk of levels by taking advantage of the fact that the things at the basic level are called the *mundane objects*. On the other hand, (Q) has the consequent infelicity: it obscures the fact that the issue is to state the population of sets of mundane objects *relative to* that of mundane objects, not to give an absolute statement of the former population. Still, I shall henceforth use (Q) as a handy formulation of the issue at hand while occasionally warning the readers against its misleading feature by talking of the *lower* realm (of mundane objects) and the *higher* realm (of sets thereof).

5.2 Reference to Set-like Objects

Most authors who hold the standard account of set give the following answer to question (Q):

(C) Any *collection* of mundane objects is a set thereof.[4]

Giving this to answer (Q) immediately invites the following question:

(Q3) What collections of mundane objects are there?

But it is hard to see how this question can be answered by the proponents of (C). Surely, it draws a circle to answer that any *set* of mundane objects is a collection thereof.

One might ask why it is necessary to consider question (Q3), for example, in assessing the answer (C) to question (Q). Why does its failure to combine with an informative answer to (Q3) make it an unsatisfactory answer to (Q) at all? The reason is that answer (C), unless supplemented with an adequate answer to question (Q3), provides no clarification of the way the population of the higher realm is determined exclusively by that of the lower realm.

To see this, it is necessary to consider what a collection is. Collections are *set-like objects*. Thus any set is a collection; sets are paradigmatic examples of *set-like* objects. In particular, any set of mundane objects is a collection thereof. And a collection, be it a set or not, is a single object; to be a collection is to be some one thing, a 'unity'. Furthermore, mundane objects are not set-like. Thus, no collection occurs at the basic level of the hierarchy. Any collection of

4. See, e.g., Boolos (1971, 491), Drake (1974, 1), Shoenfield (1977, 323ff), and Moschovakis (1980, 449ff).

mundane objects, like any set thereof, is an object outside the realm of mundane objects. Consequently, (C) depends on referring to *additional* objects (viz. collections of mundane objects), which lie *outside* the lower realm, in stating the condition for the existence of things in the higher realm. The following formulation of (C) helps to see this clearly:

> If there is a *collection* y that has only mundane objects as members, then there is a set x that is identical with y.

The antecedent of the answer (C) in this formulation has a quantifier the range of which includes some non-mundane objects, viz. collections. So it states the population of the higher realm in terms of that of the 'third' realm, *as well as* that of the lower realm, and disregards the idea that this population *alone* suffices to determine the population of the higher realm.

Now, it should be clear why satisfying the following is a desideratum for an answer to (Q):

> *Desideratum of Local Reference*: An answer to question (Q) must state the condition for the existence of sets of mundane objects with exclusive reference to mundane objects.

Answer (C) obviously violates this desideratum, which subjects it to the objection raised above. By contrast, an answer that satisfies the desideratum settles question (Q) directly–without raising a further question, such as (Q3). The answer (N) that I give to question (Q) has this merit. To explain this, it is necessary to elaborate on its content.

5.3 Plural Reference to Mundane Objects

Recall the answer:

> (N) *Any number of* mundane objects *form* a set.

Notice that we can reformulate it as follows:

(N') If there are *some* mundane objects, *they* (i.e., those mundane objects) *form* a set.

Now, to clarify the content of the answer, it is necessary to explain the meanings or functions of the italicized phrases in (N) or (N').

First, the predicate " . . . form —". To see what I mean by the predicate, consider the following true statements in which it occurs:

2 forms {2}.
3, 9 and 17 form {3, 9, 17}.
The natural numbers form the set *N* of natural numbers.
The integers that Pythagoras liked form a set.
If Tom's favorite numbers are 9 and 12, they form {9, 12}.

The following statements, by contrast, are false:

2 forms {1}.
1 forms {1, 2}.
1 and 2 form {1}.

Thus, we can define the predicate as follows:

. . . form (or forms) *y* if, and only if, *y* has exactly (i.e., all and only) . . . as members.

Now, notice that the predicate has a special character. Its first argument place, marked by " . . . ", may admit plural terms (e.g., "3, 9 and 17", "they", or "the natural numbers") *as well as* singular terms (e.g., "1" or "it"). This feature contrasts the argument place with the more familiar ones marked by the familiar variables "*x*", "*y*", etc., such as those marked by "*x*" or "*y*" in "the successor of *x*" (in symbols, "S(*x*)") or "*x*+*y*". These argument places, which admit singular terms, do not admit plural terms. Now, the first argument place of " . . . form—"

is to these familiar argument places roughly what plural terms (e.g., "they")[5] are to singular terms (e.g., "it"). Thus I shall use plural counterparts of the familiar variables, such as the plural counterpart "*xs*" of "*x*", as variables marking such special argument places (as in "*xs* form *y*"). Accordingly, I shall say that "*x*" and "*xs*", for example, are *singular* and *plural variables*, respectively.

Now, it is straightforward to see how the expressions "*some*" and "*any number of*" in (N) or (N') function. The function of their singular counterparts, such as the expressions "any" and "a" in the sentences below, are more commonly noticed:

(1) *Any* natural number is an integer.

(2) If there is *a* natural number, it is an integer.

They occur in these sentences as quantifiers binding singular variables (e.g., "*x*") or singular pronouns (e.g., "it" in (2)). "Any number of" in (N) and "some" in (N') are like "any" in (1) and "a" in (2), respectively, *except that* they bind plural variables (e.g., "*xs*") or plural pronouns (e.g., "they" in (N')). So we can say that they are *plural quantifiers* whereas "any" and "a" are *singular quantifiers*.

5.4 Exclusive Reference to Mundane Objects

We can now consider how answer (N)[6] satisfies the Desideratum of Local Reference. Notice that the answer can be reformulated as follows:

(N) If there are some ***ms*** such that ***ms* are mundane objects**, there is a set *x* such that ***ms* form *x***,

where "*ms*" is the plural counterpart of the restricted singular variable "*m*" that ranges over mundane objects. To satisfy the desideratum, no term in the expressions in boldface, except the variable "*x*" in "*ms* form *x*", can refer to (or

5. More precisely, it amounts to such mixed terms as "he, she, it or they".

6. Henceforth, I shall not distinguish (N) from (N') or other equivalent formulations.

range over) things other than mundane objects. Thus (N) satisfies the desideratum if the plural variable "*ms*", the only term occurring therein except "*x*", refers to nothing outside the realm of mundane objects. I maintain that the variable meets this condition.

To see this, consider, for example, the following statement:

(3) Some children sing together.

This can be paraphrased using the plural variable "*ps*" whose singular counterpart (i.e., "*p*") ranges over people:

(3') Some *ps* are such that *ps* are children and sing together.

To take the view that the plural variable "*ps*", for example, refers to something of a sort (be it a set, collection, aggregate, etc.) *over and above* people is to maintain that the plural existential generalization (3) or (3') asserts the existence of one (or more) of the additional things. To say, for example, that "*ps*" refers to a set of people (over and above people) is to say that (3) or (3') is logically equivalent to a statement of the form "There is a (non-empty) set *x* of people such that . . . *x* . . .", which implies the following:

(4) There is something (e.g., a set) of which something (e.g., a person) is a member.

Thus, on the view in question, (3) implies (4), which asserts the existence of a set, collection, etc. But this is not correct. To see this, consider the following:

(5) Tom and Julie are children that sing together.

It should be clear that

(A) Statement (5) implies statement (3).

Next, I shall argue that

(B) Statement (5) does *not* imply statement (4).

This thesis, together with Thesis (A), implies the desired conclusion that (3) does not imply (4). Now, to see that Thesis (B) holds, consider the following statement:

(6) Tom is a child, Julie is a child, and Tom sings (together) with Julie.

It should be clear that this statement does not imply statement (4). But it implies statement (5); it implies "Tom and Julie are children and Tom and Julie sing together." Thus (5) does *not* imply (4).

This completes my argument against the view that a plural variable (or quantifier) refers to something over and above the things that its singular counterpart ranges over. In particular, the plural variable "*ms*" does not refer to a collection of mundane objects. Thus answer (N) satisfies the Desideratum of Local Reference.

5.5 Concluding Remarks

Recall the two answers to (Q) that I have contrasted:

(N) *Any number of* mundane objects form a set.
(C) *Any collection of* mundane objects is a set.

The critical difference between them is this: "Any" and "collection" in (C) are used as a singular quantifier and a general term for a kind of objects, respectively; by contrast, "any number of" in (N) is used as a plural quantifier (which does not contain "number" as an independent semantical unit). The quantifier that occurs in (N), which can be seen to bind the restricted plural variable "*ms*", the plural counterpart of the variable "*m*" that ranges over mundane objects, does not involve commitment to the existence of anything other than mundane objects. Thus (N) refers exclusively to mundane objects in stating the condition for the existence of sets.[7] This feature of (N) is the key to its relation to the idea of population determination. Let me add that there is more to the answer. The statement of the condition just appeals to the fact that mundane objects exist, not to the ways they exist. Only the *existence* of Tom and Julie as mundane objects is relevant to the existence of, e.g., the doubleton {Tom, Julie} as a set of mundane objects. It does not matter whether they are children or adults, material or immaterial, or whether they sing together, etc. I think this helps to clarify the idea that the population of sets of mundane objects is determined exclusively by that of mundane objects. This idea is a central component of the notion of set, according to which a set is something determined by its members. Thus I think (N) is a statement very pertinent to what a set is.

---

7. Recall the general question (Q1) concerning the relationship between the population of a given level in the hierarchy of sets and those of the levels below it. Needless to say, a straightforward generalization of (N) can be given as an answer to (Q1):

> *Any number of* things occurring below a given non-basic level form a set at that level.

But we cannot go a step further and maintain:

> (N') *Any number of* objects form a set.

This is inconsistent, leading to a version of Russell's paradox–assuming that there are at least two objects or that there is the empty set. In any case, all of the objects are *a number of* (i.e., *some*) objects, but there is no set with all of them as members. For this reason, not for idle pedantry, I have raised the question of population determination only in the context of levels of the hierarchy. Notice also that, on my treatment, the falsity of (N') does not require the existence of a set-like object that is not a set (e.g., a *proper* class or collection) so that we can assert the negation of (N') without committing ourselves to the existence of such an object. Clearly, this is the right result.

Bibliography

Almog, Joseph

1991 "The what and the how", *Journal of Philosophy*, 88: 225-44.

Barwise, Jon (ed.)

1977 *Handbook of Mathematical Logic* (Amsterdam: North-Holland Publishing Co.).

Benacerraf, Paul

1965 "What numbers could not be", *Philosophical Review*, 74: 47-73; reprinted in Benacerraf and Putnam (eds.) (1983): 272-94.

1973 "Mathematical truth", *Journal of Philosophy*, 70: 661-80; reprinted in Benacerraf and Putnam (eds.) (1983): 403-20.

1982 "Skolem and the skeptic, I", *Proceedings of the Aristotelian Society, Supplementary Volume*, 59: 85-115.

Benacerraf, Paul and Hilary Putnam (eds.)

1983 *Philosophy of Mathematics*, 2nd ed. (Cambridge, Cambridge University Press).

Bernays, Paul

1935 "Sur le platonisme dans le mathématiques", *L'Enseignement mathématique, 1st series*, 34: 52-69; translated as "On platonism in mathematics" by C. D. Parsons in Benacerraf and Putnam (eds.) (1983): 258-71.

1946 "Review of K. Gödel's 'Russell's mathematical logic'", *Journal of Symbolic Logic*, 11: 75-79.

Bigelow, John

1988 *The Reality of Numbers* (Oxford, Clarendon University Press).

Black, Max

1971 "The elusiveness of sets", *Review of Metaphysics*, 24: 614-36.

Bocheński, I. M. et. al. (eds.)

1956 *The Problem of Universals: A Symposium* (Notre Dame, Ind.: University of Notre Dame Press).

Boolos, George

1971 "The iterative conception of set", *Journal of Philosophy*, 68: 215-32; reprinted in Benacerraf and Putnam (eds.) (1983): 486-502.

1975 "On second-order logic", *Journal of Philosophy*, 72: 509-37.

1984 "To be is to be a value of a variable (or to be a value of some variable)", *Journal of Philosophy*, 81: 430-49.

1985a "Nominalist platonism", *Philosophical Review*, 44: 327-44.

1985b "Reading the *Begriffsschrift*", *Mind*, 94: 331-44.

Browder, F. E. (ed.)

1974 *Proceedings of Symposia in Pure Mathematics: Vol. 28* (Providence, RI: American Mathematical Society).

Burge, Tyler

1977 "A theory of aggregates", *Noûs*, 11: 97-117.

Butts, Robert E. and Jaakko Hintikka (eds.)

1977 *Logic, Foundations of Mathematics and Computability Theory* (Dordrecht-Holland: D. Reidel Publishing Co.).

Cantor, Georg

1883 "Über unendliche, lineare Punktmannigfaltigkeiten, 5", *Mathematische Annalen*, 21: 545-86; reprinted in Cantor (1903): 165-209.

1890/1 "Über eine elementare Frage der Mannigfaltigkeitslehre", *Jahresbericht der Deutschen Mathematiker-Vereinigung*, 1 (1890-91): 75-8; reprinted in Cantor (1903): 278-80.

1895-7 "Beiträge zur Begründung der transfiniten Mengenlehre" (Parts 1 and 2), *Mathematische Annalen*, 46 (1895): 481-512 and 49 (1897): 207-46; reprinted in Cantor (1903): 282-356; translated as Cantor (1915).

1899 "Cantor an Dedekind", in Cantor (1903): 443-7 & 451; translated as "Letter to Dedekind" in van Heijenoort (ed.) (1967): 113-7.

1903 *Gesammelte Abhandlungen: mathematischen und philosophischen Inhalts*, edited by Ernest Zermelo (Berlin: Springer); reprinted (Hildescheim: Olms, 1962).

1915 *Contributions to the Founding of the Theory of Transfinite Numbers*, translation of Cantor (1895-7) by P. E. B. Jourdain (Chicago and London: Open Court); reprinted (New York: Dover, 1952).

Chang, C. C. and H. Jerome Keisler

1990 *Model Theory* (North-Holland, Elsevier), 3rd ed.

Copi, I. M. and J. A. Gould (eds.)

1967 *Contemporary Readings in Logical Theory* (New York: Macmillan).

Davidson, Donald

1967 "The logical form of action sentences", in Rescher (ed.) (1967): 81-95; reprinted in Davidson (1980): 105-21.

1980 *Essays on Actions and Events* (Oxford: Clarendon Press).

Dedekind, Richard

1888 *Was Sind und Was Sollen die Zahlen*? (Brunswig: Vieweg); translated as "The nature and meaning of numbers" in Dedekind (1901): 29-115.

1901 *Essays on the Theory of Numbers*, translated by W. W. Beman (Chicago, Open Court); reprinted (New York: Dover, 1963).

Dummett, Michael A. E.

1991 *Frege: Philosophy of Mathematics* (Cambridge, Mass.: Harvard University Press).

Drake, Frank R.

1974 *Set Theory: An Introduction to Large Cardinals* (Amsterdam, New York: North Holland Publishing Co.).

Enderton, Herbert B.

1972 *A Mathematical Introduction to Logic* (New York, NY: Academic Press).

Fine, Kit

1981 "First-order modal theories: sets", *Noûs*, 15: 177-205.

1994 "Essence and Modality", *Philosophical Perspectives, Vol. 8: Logic and Language*: 1-16.

Frege, Gottlob

1884 *Die Grundlagen der Arithmetik* (Breslau: Köbner).

1894 "Über Begriff und Gegenstand", *Zeitschrift für Philosophie und Philosophische Kritik*, 100: 25-50; translated with additions as "On concept and object" in Frege (1979): 87-117.

1950 *The Foundations of Arithmetic* (Oxford: Basil Blackwell); translation of Frege (1884) by J. L. Austin; 2nd revised edition (Oxford: Basil Blackwell, 1980).

1979 *Posthumous Writings*, Hans Hermes et. al. (ed.) (Oxford: Basil Blackwell).

Gödel, Kurt

1947/64 "What is Cantor's continuum problem?", *American Mathematical Monthly*, 54 (1947): 515-25; a revised and expanded version in Benacerraf and Putnam (eds.) (1983): 470-485.

Goodman, Nelson

1956 "A world of individuals", in I. M. Bocheński et. al. (eds.) (1956): 13-31; reprinted in I. M. Copi and J. A. Gould (eds.) (1967): 204-16.

Grandy, Richard E.

1973a "Reference, meaning, and belief", *Journal of Philosophy*, 70: 439-52.

1973b "Comments on Moravcsik's paper", in J. Hintikka, J. Moravcsik, and P. Suppes (eds.) (1973): 295-300.

1976 "Anadic logic and English", *Synthese*, 32: 395-402.

1977 *Advanced Logic for Applications* (Dordrecht, Holland: D. Reidel Publishing Company).

Hallett, Michael

1984 *Cantorian Set Theory and Limitation of Size* (Oxford University Press: Oxford).

Hart, William D.

1975/6 "The potential infinite", *Proceedings of the Aristotelian Society* (*New Series*), 76: 247-64.

Hazen, Allen P.

1993 "Against Pluralism", *Australasian Journal of Philosophy*, 71: 132-44.

Hintikka, J., J. Moravcsik, and P. Suppes (eds.)

1973 *Approaches to Natural Language* (Dordrecht, Holland: D. Reidel).

Jech, Thomas (ed.)

1974 *Axiomatic Set Theory, Proceedings of Symposia in Pure Mathematics, Vol. 13, Part 2* (Providence: American Mathematical Society).

Kalish, Donald et. al.

1980 *Logic: Techniques of Formal Reasoning*, 2nd ed. (New York, NY: Harcourt Brace Jovanovich).

Kenny, Anthony

1963 *Action, Emotion and Will* (London: Routledge & Kegan Paul; New York: Humanities Press).

Kessler, Glenn

1980 "Frege, Mill, and the foundations of arithmetic", *Journal of Philosophy*, 77: 65-79.

Kim, Jaegwon

1981 "The role of perception in a priori knowledge", *Philosophical Studies*, 40: 339-54.

Kripke, Saul

1980 *Naming and Necessity* (Cambridge, Mass.: Harvard University Press).

Lavine, Shaughan

1992 "Review of Maddy, Realism in Mathematics", *Journal of Philosophy*, 89: 321-6.

1994 *Understanding the Infinite* (Cambridge, MA: Harvard University Press).

Leonard, Henry S. and Nelson Goodman

1940 "The calculus of individuals and its uses", *Journal of Symbolic Logic*, 5: 45-55.

Levin, Michael

1992 "Still a horse-race", *History and Philosophy of Logic*, 13: 111-4.

Lewis, David

1991 *Parts of Classes* (Oxford: Basil Blackwell).

Lowe, E. J.

1993 "Are the natural numbers individuals or sorts?", *Analysis*, 53: 142-6.

Maddy, Penelope

1980 "Perception and mathematical intuition", *Philosophical Review*, 89: 1-22.

1981 "Sets and numbers", *Noûs*, 15: 494-511.

1983 "Proper classes", *Journal of Symbolic Logic*, 48: 113-39.

1988 "Believing the axioms, I", *Journal of Symbolic Logic*, 53: 481-511.

1990 *Realism in Mathematics* (Oxford: Oxford University Press).

Martin, Donald A.

1970 "Review of W. V. Quine, *Set Theory and its Logic*", *Journal of Philosophy*, 67: 111-4.

1974a "Hilbert's first problem: the continuum hypothesis", in Browder (ed.) (1974): 81-92.

1974b "Sets versus Classes", unpublished manuscript.

Massey, Gerald J.

1976 "Tom, Dick, and Harry, and all the king's men", *American Philosophical Quarterly*, 13: 89-107.

Mill, John Stuart

1843 *A System of Logic*, people's edition (London: Longmans, Green and Co., 1884).

Moschovakis, Yiannis

1980 *Descriptive Set Theory* (Amsterdam: North-Holland Publishing Co.).

Oliver, Alex

1994 "Frege and Dummett are two", *Philosophical Quarterly*, 44: 74-82.

Parsons, Charles D.

1974 "Sets and classes", *Noûs*, 8: 1-22; reprinted in Parsons (1983): 209-20.

1977 "What is the iterative conception of set?", in Butts and Hintikka (eds.) (1977): 335-67; reprinted in Benacerraf and Putnam (eds.) (1983): 503-29.

1983a *Mathematics in Philosophy* (Ithaca, New York: Cornell University Press).

1983b "Sets and Modality", in Parsons (1983a): 298-341.

Perry, John

1978 "Relative Identity and Number", *Canadian Journal of Philosophy*, 8: 1-14.

Pollard, Stephan
1986 "Plural quantification and the iterative conception of set", *Philosophy Research Archives*, 11: 579-87.
Pozsgay, Lawrence
1971 "Liberal intuitionism as a basis for set theory", in Scott (ed.) (1971): 321-30.
Putnam, Hilary
1967a "Mathematics without foundations", *Journal of Philosophy*, 64: 5-22; reprinted in Putnam (1975): 43-59.
1967b "The thesis that mathematics is logic", in Schoenman (ed.) (1967); reprinted in Putnam (1975): 60-78.
1975 *Mathematics, Matter and Method: Collected Philosophical Papers, Vol. I* (Cambridge: Cambridge University Press, 1975).
Quine, Willard V.
1937 "New foundations for mathematical logic", *American Mathematical Monthly*, 44: 70-80; reprinted with "Supplementary remarks" in Quine (1953): 80-101.
1948 "On what there is", *Review of Metaphysics*, 2; reprinted in Quine (1953): 1-19.
1953 *From a Logical Point of View* (Cambridge, Mass.: Harvard University Press); 2nd edition (New York: Harper & Row, 1961).
1960a *Word and Object* (Cambridge, Mass.: The MIT Press).
1960b "Variables explained away", *Proceedings of the American Mathematical Society*, 104: 343-7; reprinted in Quine (1966): 227-35.
1966 *Selected Logic Papers* (New York: Random House).
1970 *Philosophy of Logic* (Englewood Cliffs, NJ: Prentice-Hall).
1972 *Methods of Logic*, 3rd ed., the 1st and 2nd printings (New York: Holt, Rinehart and Winston).
1973 *The Roots of Reference* (La Salle, Ill.: Open Court).
1974 *Methods of Logic*, 3rd ed. (London: Routlege & Kegan Paul, 1974); reprint of the 3rd and later printings of Quine (1972).

1982 *Methods of Logic*, 4th ed. (Cambridge, Mass.: Harvard University Press).

Rescher, Nicholas (ed.)
1967 *The Logic of Decision and Action* (Pittsburgh, PA: The University of Pittsburgh Press).

Resnik, Michael D.
1988 "Second-order logic still wild", *Journal of Philosophy*, 85: 75-87.

Russell, Bertrand
1903 *Principles of Mathematics* (Cambridge: Cambridge University Press); 2nd ed. (New York: W. W. Norton & Co., 1937).
1919 *Introduction to Mathematical Philosophy* (New York, Macmillan); 2nd ed. (London: George Allen and Unwin, 1920).

Schoenman, R. (ed.)
1967 *Bertrand Russell, Philosopher of the Century* (London: Allen and Unwin).

Scott, Dana S.
1974 "Axiomatizing set theory", in Jech (ed.) (1974): 207-14.

Scott, Dana S. (ed.)
1971 *Axiomatic Set Theory: Proceedings of Symposia in Pure Mathematics, Vol. 13, Part 1* (Providence: American Mathematical Society).

Sharvy, Richard
1980 "A more general theory of definite descriptions", *Philosophical Review*, 89: 607-24.

Shoenfield, Joseph R.
1977 "Axioms of set theory", in Barwise (ed.) (1977): 321-44.

Simons, Peter M.
1982a "Number and manifolds", in Smith (ed.) (1982): 160-98.
1982b "Plural reference and set theory", in Smith (ed.) (1982): 199-260.
1982c "Against the aggregate theory of number", *Journal of Philosophy*, 79: 163-7.

Skolem, Thoralf
1923 "Einige Bemerkungen zur axiomatischen Begründung der Mengenlehre", *Matematikerkongressen i Helsingfors den 4-7 Juli 1922, Den femte skandinaviska matematikerkongressen, Redogörelse* (Helsinki, Akademiska Bokhandeln): 217-32;

translated as "Some remarks on axiomatized set theory" in van Heijenoort (ed.) (1967): 290-301.

Smith, Barry (ed.)
1982 *Parts and Moments: Studies in Logic and Formal Ontology* (München: Philosophia Verlag).

Stenius, Erik
1974 "Sets", *Synthese*, 27: 161-88.

van Aken, James
1986 "Axioms for the set-theoretic hierarchy", *Journal of Symbolic Logic*, 51: 992-1004.

van Heijenoort, Jean (ed.)
1967 *From Frege to Gödel* (Cambridge, Mass.: Harvard University Press).

Wright, Crispin
1982 "Skolem and the skeptic, II", *Proceedings of the Aristotelian Society, Supplementary Volume*, 59: 117-37.

Wang, Hao
1957 "The axiomatization of arithmetic", *Journal of Symbolic Logic*, 20: 145-157.
1974a *From Mathematics to Philosophy* (Routledge and Kegan Paul)
1974b "The concept of set", in Wang (1974a): 181-223; reprinted in Benacerraf and Putnam (eds.) (1983): 530-70.
1977 "Large Sets", in Butts and Hintikka (eds.) (1977): 309-33.

Zermelo, Ernest
1930 "Über Grenzzahlen und Mengenbereiche", *Fundamenta Mathematicae*, 16: 29-47.

Index

www.ingramcontent.com/pod-product-compliance
Ingram Content Group UK Ltd.
Pitfield, Milton Keynes, MK11 3LW, UK
UKHW021958180926
475701UK00003B/78

* 9 7 8 0 4 1 5 8 6 6 8 8 0 *